100,000 Birthdays

A novel about all of us

Cynthia Rogerson

100,000 BIRTHDAYS © 2025 CYNTHIA ROGERSON

Cynthia Rogerson asserts the moral right to be identified as the author of this work in accordance with the Copyright, Designs and Patents Act 1988

This is a work of fiction. Names, characters, places and incidents are either the result of the author's imagination or are used fictitiously.

All rights reserved. No part of this publication may be reproduced, stored in or introduced into a retrieval system or transmitted in any form or by any means (electronic, mechanical, photocopying, recording or otherwise) without the prior written permission of the author.

Isbn: 978-1-914399-43-5

This book is sold subject to the condition that it shall not be resold, lent, hired out or otherwise circulated without the express prior consent of the author.

Printed and bound in Great Britain by Clays Ltd, Elcograf S.p.A

Cover Design © 2025 Mercat Design, images courtesy of Dreamstime. All Rights Reserved

SPARSILE BOOKS

Praise for the book

"100,000 birthdays is a major feat of the imagination - a unique concept and a thought-provoking read."
SG MacLean, Bookseller of Inverness

"A witty and brilliant ride through time and space as one of the best Scottish novelists of recent years spreads her wings."
Roger Hutchinson, Calum's Road

"100,000 birthdays is an audacious exploration of one life and of all life, ever. Playful, surprising, profoundly thoughtful, supremely enjoyable and often very moving this is a memoir-cum-history-cum-novel that you will never forget."
Sean Lusk, The Second Sight of Zachary Cloudesley

"A wonderful book, highly original, often profound, very funny and written with so much heart."
Doug Johnstone, The Space Between Us

"Joyful, exuberant, outrageously audacious, Cynthia Rogerson's genre-busting new work is a beautiful exercise in wonder. This is a book that ponders the wild improbability of any human existence. Not quite a memoir, not quite a novel, almost but not quite science and spanning whole aeons of existence in a series of well crafted chapters, 100,000 Birthdays will make your head spin. It will also make you laugh, which is a rare joy these days I find."

Stephen May, Green Ink

*This book is dedicated to my (roughly)
four quad-trillion ancestors.
Thanks guys.*

Not one of your pertinent ancestors was squashed, devoured, drowned, starved, stranded, stuck fast, untimely wounded, or otherwise deflected from its life's quest of delivering a tiny charge of genetic material to the right partner at the right moment in order to perpetuate the only possible sequence of hereditary combinations that could result—eventually, astoundingly, and all too briefly—in you.

> Bill Bryson (A Short History of Nearly Everything)

My suspicion is that the universe is not only queerer than we suppose, but queerer than we *can* suppose.

> JBS Haldane (Possible Worlds)

There's an old joke. Two elderly women are at a Catskill mountain resort, and one of them says, 'Boy, the food at this place is really terrible.' The other says, 'Yeah, and such small portions.' Well, that's essentially how I feel about life – full of loneliness and misery and suffering and unhappiness, and it's all over much too quickly.

> Woody Allan (Annie Hall)

Preface

When John Keats was dying in his tiny Roman apartment with the view of the Spanish steps, he wrote a letter to his friend Charles Brown (good grief!) in England: *I have an habitual feeling of my real life having passed and that I am leading a posthumous existence...however, I will not speak on that subject.*

I feel the same, only I seem to be leading a pre-humous existence and I *am* speaking on the subject. Every evening at dinner I raise a glass to Kevin, the first life form, because I am at least .000000000000000003% made of him (as are you, and you, and you too). I always experience a jolt of affectionate familiarity, thinking of Kevin, as if I'm there at the start too. Mute and microscopic, but there. Bobbing in a young not-very-salty sea.

What was here before Kevin? I had a look last week and I'm pretty sure (no one knows with certainty) this is what I saw: There was nothing in the universe because there was no universe. Just a speck into which was compressed everything that would later become all the stars and planets and asteroids and meteorites. That would become Earth and us. It was—our beginnings were—squeezed into such a small space that when it suddenly exploded it was like a very large man who'd been saving up his sneeze until he could really let rip. Out it all came, and within minutes the three elements essential for Life formed: hydrogen, carbon and a tiny bit of helium. Carbon being king, of course. Pretty much everything boils down to carbon.

Then five or so billion years ticked by. Life was killing Time somewhere, maybe flitting from star to star singing *la la la*, not doing anything much. Matter was flying or floating around, some was burning, some was cold. Earth wasn't Earth yet, but it wasn't dark or particularly cold because the sun was already here. It was the new kid on the block, just arrived but already making the rules. Very bold, very brassy, very much the centre of attention. Still, overall it was peaceful where I was floating—a pocket of sunny space Earth would occupy soon. Lonely I suppose, because not only was no one home, there was no home. I dog-paddled for a while in that warm sunny space, thinking these thoughts, and then noticed the sound of my heart beating. Ba boom, ba boom, ba boom. After a while I fell asleep to it. I dreamed about nice things happening. I was fourteen and a cute boy smiled at me. I was fifty-six and found those tortoiseshell glasses I thought I'd lost. And then some craziness woke me. Wind, noise, fragments of matter flying everywhere.

Whap! Bang! Watch out!

But within seconds, it was eerily calm again as Earth slowly and silently, almost gracefully, began to exist. She reminded me of a very hungover woman waking up somewhere strange and slowly remembering her own boundaries. Tiptoeing around, wondering where she left her bag and shoes, where she lived, what her name was. Yes, the sneeze explosion was a long time ago, but I don't think it felt like that to her—it felt like it happened last night. She survived by retrieving all her loose bits and gathering them up, pulling them into herself. By hunkering down, closing her eyes and trying to hold herself together while being hit on the head over and over by asteroids. On top of that, erupting volcanoes. Time passed for her, and after six million years or so—*Kevin*. And three and a half billion years after that, give or take a millennium? *Me*.

Me

August 14th, 1954
A millisecond before the hour if all Life is
squeezed into sixty minutes
Seattle, Washington

Let me begin by saying that normally, on a typical day, I've zero desire to relive my past. None! It's too unsettling how wobbly my perspective is. Nevertheless, here we are. It's my first birthday and we're in the most northern westerly place in America, not counting Alaska. Mercer Island is in Washington Lake, and the residents tend to work in Seattle. Being the 1950s, the commuters are mostly male, and the cost of houses means these men wear suits and fedoras. There are cheaper places to rent on the mainland, but my parents are enamoured by the word island. I need to squint and really focus, because at first the visual reception is hazy and there's a slight delay in sound, like an echo of an echo. Like trying to talk to someone on the phone when they're sitting across the room from you. The radio is on in the kitchen, which is also the living room and bedroom. Rosemary Clooney is singing *Hey There* again. All summer long that song has been playing, but my mother isn't sick of it. She sings along, something about stars being in someone's eyes, and how love's making a damn fool of them even though they used to be pretty smart.

She's wearing a dress she sewed before she was pregnant with me. Short-sleeved, light grey chequered cotton, un-ironed because she's not going out. It hangs just below her adorable knees. Her legs are bare and she's barefoot. Look at my pretty mother. Already, she has her tiny waist again. Her black hair is in a glossy bob, with high bangs and little curls in front of her ear lobes. She's singing off key, frosting a cake, and it's my birthday cake. It's morning, so although it's August it isn't hot yet. The sun flows into the room, falling in lemony shafts on everything. Because there's a tree outside and a breeze, the light keeps shifting. The cupboards are in shadow, then they're bright. My mother's chequered dress looks dingy grey, then shiny silver. My father left for class an hour ago and I cried when he went. Why wouldn't I? I will never see him again. Never! Never! Never! But I'm over it now.

There I am.

Holding on to the low coffee table, where my mother now puts the frosting bowl for me to lick. These things—my mother's singing, the creamy sugar in my mouth, the exciting sunlight—are very nice. If I was a cat, I'd be purring. Maybe some of me *is* made up of cat, or I'm in debt somehow to a cat. An ancestor who ate cat to survive, or who ate a potato which drew nourishment from the rotting corpse of a cat. This is not only possible, it is likely.

Check me out, all 28 trillion (or 75 trillion, depending on what you read) cells of me. I'm on the chubby side and most obviously composed of my recent ancestors. High Jones forehead, wispy Swarts strawberry blonde hair, enormous Nicks cornflower eyes, Vietti giggle. How many of these baby cells are with me today? Almost none. I have a life span, and so do almost all my cells. Most white blood cells last a few days before biting the bullet. Gut cells last about a week. Red blood cells croak after 120 days, while skin cells only live three weeks. Bones are replaced every ten years or so. Eggs are the biggest cells and the only visible ones. Overall, at different rates, we become almost entirely new models of ourselves every seven years—or ten, again depending on what you read. Think you have a stick-in-the-mud life? Not on a cellular level. But some cells, it turns out, are permanent. Conveniently for non-scientists with poor memories like me, they're called permanent cells. Some skeletal muscle cells stick around. I still have 60% of the heart muscle cells I was born with. And brain cells regenerate exceedingly slowly, if at all. Quite a lot of my hippocampus and cerebral cortex neurons are present right now in my one-year-old head, ticking away. And the cells in my baby retina—I'm using some of them today to look at them. (*Dude, you haven't changed a bit! Really? Thanks. Shame about the crow's feet.*) Skin cells age quickest, even though they're decades newer than retina cells. Go figure. Isn't it nice that your heart and mind and eyes are like loyal subjects? No wonder we have a sense of familiarity, even friendliness, when considering those aspects of our physical self. Not always, obviously—but overall, being in our own skin is so cosy we don't even think about it. On a heart and mind level, we keep ourselves (our cells) company throughout our lives. It's nice to know the incomers keep coming, but thank goodness for the locals.

Feeling a bit sorry for the bits of you that don't get much of a crack at life? Those sad sperm cells and colourful colon cells who die after five days? Don't fret, some of them will live on a while after you're gone. It can take hours and even days before all of you is kaput. Which is how forensics can pinpoint when death occurs. And what about those trillions of dead cells in our bodies, the ones that get replaced? Gone, gone, gone! Yet in another sense, all of them remain in their altered forms. Nothing disappears, it just changes—according to Antoine Lavoisier who wrote *The Law of Conservation of Mass* back in 1785. We incorporate so much we are unaware of. Our bodies are certainly a lot more environmentally friendly than most of us are. Nothing is wasted.

In a general sense the term *evolution* can be applied to everything, from cells to mammals to families. Everything evolves, even mountains and seas. My life is still evolving, as is yours. We go from two microscopic scraps of organic matter to a fully-fledged life form adapting (or not) to obstacles before dwindling down to decay and extinction. Awareness of our mortality informs many of our choices and perspectives. We might choose to avoid or hurtle towards adrenalin sports after our first tricycle accident. We might adapt to heartbreak by choosing to avoid future emotional intimacy—or we might keep racing towards love and be slaughtered by it over and over. If we're exceedingly lucky, we meet someone who is right for us, have the wisdom to realise this and then be happy for roughly the rest of our lives. Which will seem too short, while at the same time be the longest experience we will ever have. (There might be a dearth of vivid memories, for easy times tend to slide off the memory bank—but small price, right?)

Societies evolve too, of course. They are born, adapt, dwindle. Perhaps awareness of global warming leading to our extinction is doing the same for social consciousness. Perhaps not. From my perspective, it looks like we're making repeated feeble stances against extinction, but the sacrifice required to be effective means we'd have to give up holidays in Spain and cars. So I'm not hopeful. Or proud, for I'm part of the problem. I've benefited from electricity and oil every day of my life. And it's been a fairly cushy life. I belong to the dominant species on the planet, and my country is rich largely due to historic exploitation of indigenous people and foreign resources. Many species have bit the bullet due to our greed. I may rant against this, but maybe ranting itself is an activity only available to the not-hungry. Some days it feels hypocritical and crass not to enjoy what I have. I mean, look at this Mercer Island kitchen with the buttery sun streaming through the windows! Have a lick of that vanilla frosting!

~~~

'How old are you today?' asks my mother, not looking at me, putting away the flour and eggs and sugar. We've been practising this and I immediately shout:

'Bun!'

I've no idea what her question means, or my reply, but I understand it's the right answer.

How do I know? When she turns around to look at me, her eyes shine, her mouth stretches wide and I can see a lot of her teeth. But mainly I know because I have the laugh feeling, which is like being tickled by someone invisible.

'That's right! One year old. What a big girl you are now. My birthday girl!'

'Buday,' I say, and my mother squeals as if I've just recited Einstein's theory of relativity.

'You are a genius!'

This is something she will say to me 174,966 times. It makes me uneasy because I know it's not true, but I know she's not patronising me. I don't know what to do with it, except set it against my failing report cards and later my adult mistakes.

I'm not having that great a time today. No matter how many old photographs and home movies you've seen, it's disturbing to watch your own young self as well as *be* one's young self. I'm see-sawing between perspectives without warning and feel carsick. I am she, and I am also I. It's stressful, not to mention just plain weird. Bear with me. I've been carsick all my life, but learned not to vomit. On the surface there's a familiarity here, though I have no conscious memory of this apartment or of being a one-year-old. The subconscious, even before we're born, stores images and sounds and tastes, as well as our wordless impulses. Feeds them into a cerebral database whose sole job is to build a picture of the world that makes sense to us, adjusting it daily until no longer required because we've ceased being. Hence, the way I don't recognize this kitchen—but do. I look around closely, my heart pounding, and think: *Yikes!* These are the things that formed my first impression of the world? This corny linoleum pattern, that pinned-to-the-wall Gauguin poster of topless Polynesian women, the fickle grey of my mother's dress. Her particular way of smiling, the memory of which still, four years after her death, opens up my chest like a window opening on a beautiful morning. Like this morning. That perfect summer air on my baby skin, which has only known one other August, and only two weeks of that one. The taste of powdered sugar and butter. Right away, watching myself, I know I'm getting high on it. The genesis of my sugar addiction. It's kind of cute, in a slow train-wreck way.

~~~

I find myself staring at my mother's skin, because I'm fascinated by young skin these days. Not to mention my own baby skin, which is luminous. My current skin is age-spotted and lined, and if I touch something rough, sometimes it tears like paper. I'm starting to run into friends more often at funerals than parties. Not always, but our conversations are often about operations, pension plans, buying cemetery plots with sea views, the number of decades we've known each other, and how is so-and-so coping with widowhood. It's like we've all had a terminal diagnosis at some recent doctor visit we thought was just to sort our insomnia, but we're not panicking. We're almost jaded. Of course I've imagined dying. Some days it's an easy graceful final sigh while wearing my most flattering nightgown. But I'll probably exit while wearing something tatty, with a flourish of nonsensical words (some of them swears), flailing arms and ugly facial expressions. Possibly loud farts and disgusting wet wheezes. Like the time when I was ten and almost drowned in the McNear's Beach swimming pool. It was a birthday party and everyone else was laughing and splashing happily, but not me. I was drowning but too embarrassed to draw attention to this further proof that I was a misfit. I mean, who dies at a pool party on a sunny day? No, not a tasteful death in store for me, I think. But what can I do? Enjoy my life while I can, obviously. I'm amazed how the view from my front window looks different every day. Have clouds always been so incredible? And I finally understand why good quality cotton sheets are worth the money. When I look back at recent good times, these largely consist of things like a solid night's sleep, or binge-watching *Line of Duty* with my grumpy husband on the sofa while eating the Rees's peanut butter cups he always hides until I have a craving. (The junkie phrase *candy man* literally means candy man in his case, and I love him very much for this. Sometimes.) I notice that everything is shrinking, including me. The world used to feel vast, but now—there's only, what, seven continents? Four oceans? That's not very many. Humans live about eight decades? Why, that's hardly time to tie your shoes! Time passes so quickly and painlessly, some days it thins me, stretches me, and it's hard to stay myself. I worry I'm fragmenting, becoming less solid. Maybe this will continue until I am nearly transparent and frayed to the point of disintegration. Maybe one day I'll put on my clothes and they'll slip to the floor, leaving me naked. But I'll be invisible by then, so it won't matter. Oh, how freeing that'll be!

~~~

'Happy birthday….uh, you,' I whisper.

I find I can't say her/my name out loud. It's a linguistic nightmare. No response. I crouch down and whisper again.

'Happiest of birthdays. You won't believe what's going to happen to you.'

This is a dumb thing to say! *No referencing the future.* Luckily, she gives no sign of being aware of me. She has a yellow frosting moustache from licking the spoon. I'd say she's beautiful, but that would sound vain. And also be a lie. She's beautiful, but only in the way all one-year-olds are. Secreted in the follicles inside her ovaries are all the eggs she will ever have, including the four that will make her children, who will then make her grandchildren. On and on, like Russian dolls. No one knows how long DNA lives, because so far we can't find any evidence that it dies—though it isn't alive in the same sense we are. It survives tens of thousands of years, longer than any artefact left by any living creature. It's found at the bottom of lightless seas, inside icebergs frozen for millennia, scraped off the bones of animals that died a billion years ago. It predates anything we can imagine, and will outlast us, I suppose...once it finds a more conducive host.

~~~

'Have you pooped, honey pie?' my mother asks. She says this as if my poopy diaper would make her day because it's my poop and no one else's. That's how much she loves me.

'No!' I cry, and smile toothlessly. 'No poo!'

'Oh, you clever, clever cookie!' This is because I've strung two words together. She doesn't care that I've lied.

I'm advanced for my age, of course. I've had constant attention from both parents, and I've been walking for a month. Well, sort of walking. I can lurch for two steps before falling on my bottom. They probably play down my achievements because in their social class no one likes superior people. But I'm even cleverer than they think I am. Like the narrator in Walt Whitman's Song of Self, *I'm not contained by my hat and boots.* Nor am I just one. I began twenty-one months ago. By the time I was born, I already knew lots of things. So many things I'll not remember or express but are nevertheless imbedded in my personality. The taste and feel of embryonic fluid. Not just the sound, but the thud-thud vibration of my mother's heart. The ability to flit in and out of my evolving body to check out the environs, or just glance at the thing I ride in, the thing that made me. No siree, I am definitely not contained by my hat and boots! I've been here all those months and perhaps longer in my entirety, silly! I didn't just accumulate cell by cell like my lungs or my fingernails. Goodness me. I don't think anyone one does that.

~~~

Are these really the thoughts I have about existence when I am one? Like my body is a place on Airbnb? Cosy room for rent, maximum one occupant, flexible perimeters. Heating, electricity and foot-transport included. Food and outings, not. Central to many sights worth a visit. I'd go for that.

~~~

I have a clean diaper on now, and I'm feeling good. I smell of Johnson's baby powder, which is my second favourite smell after Mommy's skin. She picks me up. Wipes my face with a corner of her apron. It tastes of powdered sugar, so that's alright. Then my song comes on the radio. *How Much is that Doggie in the Window?* I jiggle up and down in her arms, and suddenly she's dancing us around the tiny room. Holding me tight, one hand on the back of my head, just how I like it. I can lean out as far as I like, she'll not let me fall. The room is warmer, and the sunlight is becoming more buttery than lemony. I am moist. Around and around we twirl and when the song ends with a fervent wish that the doggie was indeed for sale, she kisses my forehead and puts me down. I don't know many words yet, not really, but I understand much. No thoughts with words—just feelings, memories in images, and impulses I feel in my stomach and chest and sometimes, if music is playing, in my feet. Her sounds have a shape and a mood, and the look on her face fills in the rest.

~~~

'Happy birthday, Cynthia,' I say my name out loud for the first time, even though I suspect it's pointless saying anything at all. The air cracks. Wack! Zap! Zowie!

'Wow,' she says, blue eyes wide and she looks right at me. Whispers: 'Did you feel that? Like lightening.'

'Yeah. Go figure.' I try to act casual. Pretend this is not the first time we've conversed, and also that it's not weird for a one-year-old to have so much vocabulary. 'Don't worry about it.'

'Okay.'

Pause, while we contemplate each other. Then I say the first thing that comes into my mind. 'Isn't she pretty?' I indicate my mother. 'Our mother.'

'She's not your mom.'

'But she is. You are me. I am what you will be in sixty-seven years.'

She's as wary now as a one-year-old can be. Meaning, not very. Though her face is quite serious.

'Oh please. Tell me this is a joke.'

Then I splutter a laugh—so unexpected, saliva sprays out. You'd think my brain would've warned my mouth, but I've never coped well with extended periods of solemnity.

'What's so funny?' Still no smile.

'Nothing,' I say. 'Sorry. Don't worry. Everything's going to be alright. Basically.'

'Of course it is,' she scoffs. 'I mean, look around. Look at my beautiful mom. Look at this sunlight. Want some cake later? The frosting is incredible. By the way, did you get what you wanted?'

At first I think she's referring to my trip to the grocery store earlier, but how would she know about that? (The answer is no,

I forgot toilet paper and Stork.)

'Well, pretty much, but not everything.'

'Oh, don't worry about that. No one gets everything they want.'

You're one! What do you know? I almost say. Then I think she's channelling our mother. Mom was always saying things like that. You don't have a date for the prom because your boyfriend dumped you six months ago and no one else wants you? You're flunking in every subject except English, and your breasts are almost non-existent? Don't worry, all of that is perfectly normal.

~~~

Mommy is putting red stuff from a tube on her lips and a miracle happens. Daddy comes home. He's back! He's back! He's back! He opens his mouth and out comes a lot of noise, which I roughly translate as: I'm tired. Am I still important to you? Important enough? He picks me up, says: 'Hello pumpkin pie,' gives me a kiss and a tickle. He tickles better than Mommy does. I laugh so hard, I squirt some pee out. I truly can't believe he's here again. He kisses Mommy on her mouth and she kisses him back. I wonder why I don't get the mouth kisses, then I forget about it. Mommy moves her arms and points at the cake, which is sitting on the kitchen counter. She opens her mouth and makes a stream of sounds—the shape is stretched out like a hot dog—and the noises say to my father: Yes, yes, you are important! So important!

~~~

After dinner (meat loaf), the cake candle is lit. This scene is deeply familiar from a black-and-white photo of me in front of a cake with one candle, my father looking about fifteen years old, crouched behind me. There's an ashtray on the table with half a cigarette smoking. The edges of the photo are zig zag, cut with the pinking shears my mother uses for sewing. It's disturbing but also heady, seeing that photo animated now. I laugh suddenly, and because I am the age I am, I squirt some pee into my pants too. Then I glance at my baby self to see if she suspects we're sharing the experience of damp pants. She looks right at me, hard, and there it is. Her/my conspiratorial look.

~~~

A neighbour man showing me his hand with two fingers missing and telling me never to put my hand into an zinpretrator. That's it, and I have no idea what an zinpretrator is or why his mutilated hand survived the cull of my memories. I suppose if I was a jar of mayonnaise, the list of ingredients on my label would claim I was 89% made from my first few years and I can't even remember them. Of my first birthday, I know only this for certain: The president was Eisenhower. We lived in a studio apartment, 212c Persimmon Road on Mercer Island. My mother was 23 and her waist was twenty inches. My father was a college student studying economics on the GI Bill. And based on the birthday cake photo—my father staring at the camera like he wants to kiss it—I'm guessing my parents were still in love with each other. This makes me happy. From my distance, I know this about them: Even after fifty years, they will still need to feel themselves married, to be publicly claimed—and yet marriage will confound them. They won't know why, but they will rarely make each other happy. I don't know why either. Marriage confounds me too.

~~~

I draw back and soak up the temporary simplicity of a young marriage.

'Good luck,' I whisper before my baby self fades. 'Don't forget to have fun. And watch out on the sugar front.'

'Good luck to you too,' she replies with a smile (at last!). 'And watch out on the telling dumb stories front.'

Neither of us are natural winkers, but we make an effort. Wink like there's something in our eye, or we've got a nervous tic. And then she's gone and I am alone in front of my own kitchen sink in Scotland, looking out the window at my overgrown garden and the three steps leading to the greenhouse my husband built during Covid lockdown. It rained a few minutes ago, by the looks of it, but the sun is shining and there's a luminescence to everything, even the cobwebs. Especially the cobwebs. The kitchen smells of old dog and the cyclamen I've popped in a vase.

~~~

There are so many ways to think of a life! Can it ever be summed up, outside of fiction? I find it's hard to think of mine chronologically, as a list of events. It wriggles around, jumping backwards and forwards. Looking back reminds me of the way a road never looks the same on the return journey. It seems shorter. The landmarks seem different. In fact, some days it's almost unrecognisable. You can start to think you're on a different road altogether. Similarly, these days when I revisit my past, it only vaguely resembles the memories I have of living those times. Each phase contains contradictions. My most lonely and unhappy year, the year my first marriage ended, was also the year I laughed the hardest. And the years I look back on as my happiest certainly didn't feel that way at the time. I was lurching from day to day with my four young children, sleep-deprived and distracted, full of resentment for my husband because I was convinced he had the better end of the deal. I never would've believed how much I'd miss those days—those picnics in the rain, those times of watching Pocahontas over and over, those hours of reading A.A.Milne and Thomas the bloody Tank Engine. And that dire camping trip in Nova Scotia when I was pregnant and just discovering my first husband was not really a camper or, in fact, in any way compatible with me. Now when I recall his mosquito bites and sunburnt nose and mute English misery, I just feel fondness and guilt. I was waiting for him to be someone else. And for my life (my theoretically grand exciting life) to begin properly. And now I have a second husband who is so omnipresent, it's hard to appreciate all the ways in which my life would be impoverished without him. Maybe I'm extremely happy now and don't even know it.

~~~

I was 67 when I began writing this and now I'm close to seventy. Soon I will no longer exist in my current form. (And by soon, I mean if I live to 100, I'm pretty sure it will feel too soon.) Every atom of me will remain, but in increasingly different arrangements. Some of me will certainly go into the making of something else one day, or possibly someone else. Maybe I'll get to be part of a smart boy, or a knock-out woman. Or a beetle or a newt or a blue tit. I wouldn't mind a golden retriever experience. I think about this on and off quite a bit. Bill Bryson's books have made it impossible to forget that in a genetic sense I'm the reincarnation of all my predecessors, starting with one-celled Kevin bobbing in a young not very salty sea. In varying degrees, and in a mixture unique to me, they all ride inside and pump through my veins. Their frustrated yearnings and dire fears and insecurities, their triumphs and irrational joys, their freckles and red hair, their inclination to learn slowly, sleep fitfully, enjoy walking and parenthood, fall in love too easily, fail to keep time to music even when everyone else is clapping along. Epigenetically there might even be an atom from that traumatic moment when my first ancestor, unicellular Kevin, stretched out for a yawn and split in two.

~~~

I've had a look, and so far my ancestors don't seem that impressed with me. Their opinion shouldn't matter, but it does. I want everyone to like me. Pathetic, I know! Nevertheless, I'm the end product of them all. I incline my head in their general direction and think: *It's all your fault, guys.*

Before Homo Sapiens

Kevin

Three and a half billion years ago. Four minutes past the hour if all Life is squeezed into sixty minutes
The Ocean

It's hot and damp here in Kevin-town, and on the surface nothing much is happening. The granddaddy of us all—times a zillion—isn't pulsating. He's barely bobbing under the surface of the sea. He has no gender, but to prevent clumsiness I'm calling him him. I've got a teeny birthday cake in my pocket, candle lit in a waterproof way, because it's been one year (or the equivalent of) since he came into being. I can't say *since he was born*, because that would imply parents, living things older than him—and there are none on this planet. Kevin is the start of the whole show. Of you and me. Of oak trees and hammerhead sharks and New York cockroaches and fourteen million other species living today, of which only 1.2 million have been documented. For Kevin's appearance, more than the dawn of Homo sapiens, is the Adam and Eve moment. From nothing to something.

Look at Kevin. Just look at him!

'What are you looking at?'

'You. Do you mind?'

'Kind of. What do you want?'

'Just to look at you, really. I'm a little obsessed with you.'

Privately, I think I'm seeing a bit of Grandpa Jones in him. Something about the voice. It's hard not to look for resemblances in ancestors. Maybe impossible.

'Are you nuts?' Totally Grandpa Jones.

'A little.'

A kind of quiver goes through Kevin, and he says:

'What's that smell?'

'Oh, that's me. Sorry. Time travel gives me gas sometimes. Are you cold?' I say, to change the subject.

'Why would I be cold?'

'I don't know. You look naked.'

He's got no nucleus, and his membrane is so sheer it's hardly there. A microscopic scrap with a barely discernible perimeter, very like modern day bacteria. Quite fascinating, if you squint a bit and remind yourself he is what eventually results in you. In beautiful and contrary you.

Kevin is alone in the sea. Which is an undecided body of water, even now, because gravity keeps changing its mind. Nothing's stable (and never will be). Ice is forming and cracking someplace north of here. Closer to home where Kevin lives (lives!), in a place where land almost breaches sea, steam is hissing, water is boiling. A few hours' swim from here and there's nothing but warm mud. Bubbles are bubbling all over the place, none of them to do with oxygen because there isn't any. And every now and then, good old-fashioned thunder and lightning. A cacophony, but I find it restful. Maybe it's the absence of people.

~~~

Where did Kevin come from? Everything comes from something. Some think Kevin dropped from space...and why would that be unlikely? After all, all of Earth dropped from space. Much of it from our own star, the Sun. We are literally made of stardust. (Thankyou, Joni. Also Dr Ashely King, planetary scientist who first claimed the elements in our bodies were made in a star). Then again, maybe Kevin has nothing to do with space at all. Maybe, as abiogenesis scientists believe, he's a local boy through and through, manufactured with RNA domestic materials and nature-made labour with some luck and coincidence thrown in. Or maybe he was created by God, a mysterious being with many names and for whom the concepts eternity and infinity are as comprehensible as ironing board or miles-per-gallon are to us. We can't visualise eternity or infinity because our minds naturally grope for the edges of things, for the beginnings and endings—which is why we're not God. Unless we all are, of course, as the Pantheists believe. When I was seventeen, astronauts using a sounding rocket discovered a black hole in the milky way and called it Cygnus X-1. It looked like a squashed donut and I wondered if God lived there. At that time, God was a cuddly bearded man like the folk singer Burl Ives. I was surprised this idea wasn't in the headlines. *God's House Located.* Even though I was starting to doubt his existence, a bit of me hoped God wasn't a charlatan like the wizard in Wizard of Oz. That would not have been good news for me at that age. I needed God to have a solid address and name.

'Why do you care, anyway?' asks Kevin suddenly. 'Why do you always need words for things?'

'Well, because...just because,' I answer feebly. 'Because without words, how can we talk about a thing? We have to call it something even if we don't know what it is.'

'Why?'

'So we don't mix it up with something else. I guess.' *Damn him,* I'm thinking.

Then he says in a consoling way, as if I'm the quivering unicellular being: 'Okay. God is probably an okay word for the thing we have no words for. Because—a million stories about it or not—no one really knows what Life is all about.'

Instantly I'm singing *What's it all about, Alfie?* in my head.

'No one?' I ask.

'Not a single person. Certainly not you.'

Jiminy cricket, I think.

'Or Alfie,' he adds, as if this day was not surreal enough.

'Weird stuff happens. For instance,' he continues self-importantly, 'just yesterday I woke up feeling normal, no particular plans for the day, but when I stretched for a yawn...well, look at me. By the time I finished yawning, I'd split into two. What's that about?'

Wait a minute, I think. *Kevin's not alone anymore?* Then I notice there's a replica Kevin floating about a centimetre away. In fact Kevins are suddenly coming out of the woodwork—but I discreetly say nothing even though he's looking at me expectantly.

'I mean, holy fucking moly, right?'

He sounds exactly like my brother now. Annoyed and amused and requiring a sympathetic response. I'm too tired to say anything at first. And there's the anti-climax factor. I always knew splitting was on the agenda.

'Okay. Yeah, Kevin. Must've been a shock. Do you mind if I continue addressing you as if you are not plural and genderless?'

'Suit yourself.' And just like that, for me, he's back to being singular.

'Thanks.' I have trouble with generalities and need individual narratives, but I don't tell him that. He has too little respect for me as it is.

'Whatever.'

'Thanks, I'm so grateful. I mean I...'

'I said, *whatever*,' he blurts half way through a yawn.

Perhaps Kevin sounds jaded because he has no illusions. If he had teeth, he'd be hanging on to Life by the skin of them. Maybe he gets through his days by telling himself (in his collective unself-aware way) he's waiting. Enduring. Silently splitting over and over in quiet desperation because what else is there to do? Patience is probably the primary trait required by Life at this stage, because survival depends on accepting delayed gratification. It means embedding an ambition to stay alive into your DNA, then splitting. Kevin is a huffing-and-puffing runner in a long-distance race, passing the baton a millisecond before collapsing in a heap. A race he has no hope of seeing the end of.

'The bad news, Kevin, is you're not important.'

'Great. What's the good news?'

'You're not important. Liberating, right?'

And the same, of course, applies to you and me.

~~~

The Kevins look identical, but they aren't. *Look closer*. Each is unique, as is every single living thing yet to come. Even identical twins do not have identical fingerprints. Individuals may not be important, but individuality is. Survival tip number two: Be yourself, it's the best card to play. It may not always benefit you personally, but it will assist the evolution of your species. From the start, Kevin has been experimenting with Life—or Life has been experimenting with Kevin. Whichever way you look at it, each Kevin is a survival experiment—and there's no short cut. No one knows (if there was anyone here to wonder) which Kevin will be best at keeping the Kevins in existence. A matter of trial and mostly error—but then Kevin is splitting so bountifully, it hardly matters. The Kevins that are most adaptable, that learn to repair themselves and alter their function and specialise, will really go places. Eventually, in a few billion years, the cleverest cells will be ready at the drop of a hat to become a chimpanzee's eardrum or a hair on a hare's hind leg. They'll have learned to work together for the good of the whole as well as personal survival. Society is born right now, on a cellular level.

'So tell me, what's it like, being just one cell big?'

'I don't know. It feels normal, I guess. I've always been one-celled.'

'I think you're amazing,' I gush.

'Me. Amazing,' he replies flatly.

'Where did you come from?' I ask, then hit my head with my hand. I'd told myself I wouldn't ask this. Rude, rude, rude!

'I have no idea. I wasn't here before I was here.'

The tiniest puff of air, and I convince myself Kevin has just yawned again.

'Look, I just woke up one morning alive, okay?'

'Okay. Sorry,' I say.

~~~

If I was writing a novel about Kevin, I'd begin like this:

Once upon a time, not ten feet from where we're sitting, a random carbon atom was hanging out with a common-as-muck hydrogen atom. They were just minding their own business, loitering, bumping into each other now and then and saying *excuse me*. Then after a while they began sliding neatly past each other, pretending not to notice, flirting in their star-dusty way—but both as life-less as lava. Then a bit of organic flotsam—like a stale hot dog or apple core—drifted into Earth's orbit from an asteroid, and after a while settled down to join the flirting atoms on Earth's surface. It just happened those atoms contained precisely the ingredients the flotsam needed to…to rehydrate. To materialise into something alive. What or who made the flotsam in the first place? Well, Life itself, I might state boldly. Life with a capital letter is the culprit! Also, the main character in my novel. A non-physical entity no one can visualise or imagine, whose sole and overwhelming desire is to continue *being* via any form that happens along. A kind of perennial squatter or hitchhiker. (Maybe the dead can hitch rides in those who still think of them, or simply those who leave the door open. Maybe it turns out Love really is a ticket to ride.) But what or who made Life? *I don't have a clue*, my narrator would admit. Still boldly because I don't think anyone knows, there's no shame in not knowing, and also because I think non-resolution is the essence of good literature. On and on, back and back we go, like the Beach Boys song about there being no beginning or ending, as long as you're depending on love. The last bit is nonsense of course, a twentieth-century illusion, and why should anyone need romantic love anyway? Surely it's miraculous enough, just this impersonal never-ending-ness. I'd imbue Life with a large dash of disinterested benevolence so the reader could feel sympathetic. I'd title it *The First True Adventure Story* and the subtitle might be something like: *A timeless classic, extremely personal to everyone.*

'I'm not sure about the novel version,' says Kevin.

'Why not?'

'Well, it's not very flattering to me, is it? Made of stale hot dogs and apple cores. At the behest of some magical creator.'

'Huh. If you look at it like that.'

'I do. Actually.' I think he's sniffing, in his non-nose way. Pride, I suppose. And why shouldn't he feel pride?

~~~

If I was writing the scientific version, I'd use lots of numbers and long words that are hard to spell and impossible to remember, much less pronounce. I'd say something clever along the lines of: Various factors—inanimate gases, chemicals, atoms, plus the weather that day and the temperature of the sea—combined to create the circumstances for Kevin to, uh, to *ping* into life. He is a fortuitous accident, but entirely plausible, once one has entered all the available data into the, erm, the space time continuum infrastructure thingamajig. I'd need to actually read some academic books, which are not fiction. I'm not sure I'm physiologically capable of that.

'Better, but you'll never manage it,' Kevin says. 'You can't even work your microwave.'

Jesus, I think. How can he know these things?

~~~

If I was writing the science fiction version, I might say: The ping that pinged Kevin into life, thinks it's the beginning of Life. But the thing that pinged the ping is a migratory and predatory population from planet Zook who are colonising the universe for their own wicked benefit. They're like zombies, only invisible, and once you've been colonised you don't know it. You think you're still normal. I'd give my protagonist (played by George Clooney in the movie) a time travel device cleverly disguised as an egg timer.

'Now, that's what I call a story!' Kevin says.

'Maybe. But it might give me nightmares to write it. And plot is not my forte.'

~~~

If I was writing the religious version (which might appeal to the roughly 46% of Americans who do not buy any evolution theory) it would be another excuse to capitalise Life, the same way God is capitalised as a proper noun. Like God, Life deserves respect. Also because, who knows, maybe Life *is* God. Not an individual entity with an army of angels or a figment of man's lonely and frightened imagination, but a mysterious force that propels us all forward. That, at least since Kevin but probably further back, informs the choices of every life form, all geared, needless to say, towards perpetuity. Maybe we are all instruments of Life. Hey, it's just an idea. I could also write a version in which God is exactly how the Bible paints him, and all Life is created in a week. But I probably wouldn't sell many copies because this version would be too wacky, like the Adam and Eve story going out of fashion. (Only 20% of Americans go to church weekly.) It's too unlikely, too quaint, like a fairy story—although I suppose one day a generation will look back at some of our ideas and think them quaint and unlikely too. *How silly those twenty-first-century people were*, they'll say. *They actually thought the universe came out of a tiny speck going bang, when everyone knows*…blah, blah, blah. And a thousand years from then, another generation will say the equivalent of the same thing. It's worth noting that as far back as 2000 years, and maybe as far back as 200,000 years, humans understood there was something fishy going on with their beginnings. Something that defied logic. The story of God—or indeed, Gods—wraps it up pretty nicely.

'Or goddesses,' murmurs Kevin.

'Oh yeah. Of course.'

I keep forgetting he's listening to every word whether I say it out loud or not. Which is gratifying, if a little unnerving. (Note to self: avoid unkind Kevin thoughts.)

Personally, I'm not hugely dismissive of any genesis story. What if God exists but is none of the things we've imagined? What if he's the spotty-faced bloke called Trev who's in charge of the shopping carts at Tesco? We're such snobs. Anything along these lines is possible, as far as I'm concerned. If someone had told me, aged twenty, that one day I'd carry a bit of plastic around and it would enable me to talk to anyone in the world at any time for free, as well as show me movies, give me news, take photographs and videos, allow me to know the location of anyone else in my contact list, I'd say: *Oh baloney*. I don't mention this out loud, as if that would matter to omniscient Kevin. The point being, anything is possible.

Pause, while he bobs and I think.

'So,' says Kevin finally, with a note of wariness. 'Which kind of book is this, then?'

'None of those, I guess. Not a proper novel, not science fiction, not science essays, not a religious tract. It's just my personal take on Life with a capital letter. About the life forms that end up—after three and a half billion years—in me being me. You could never be a fictional character, Kevin. I mean look at you! You're here, you really exist!'

'Well, duh.'

'I'm hoping there'll be some connection somewhere along the line between my life and Life's life. To tell you the truth, I've done some really dumb things. Maybe understanding the biological imperatives of my species, and all the species that led to my species, will help make sense of some of my mistakes. After all, every aspect of us has, or has had, a purpose. Maybe the cycles that occur on a cellular level, are no different from cycles that occur on a whole body of cells. Or the length of day. Or the seasons of the year. Maybe it's all part of the same thing.'

Long pause, and I wonder if I've been talking too long. Is he asleep?

'I have no idea what you're talking about.' He spaces these words out, emphasizing *no idea*.

Christ. It's so hard, knowing how to talk to the thing you come from.

'I guess this book is an autobiography, Kevin. In the broadest cellular sense.'

'Is it true?'

'All autobiographies are highly subjective, but don't worry. I've painted you in a really good light. You look great. Readers are going to love you.'

'Really? You think so? You really think so?'

Quite sweet, the way he responds to flattery. Makes me

wonder if he's ever had a compliment. I know he's never paid one, because who'd be around to receive it?

'Truly. You'll have a fan club. Everyone loves a mystery.'

'I don't feel mysterious. I'm a one-cell dude.'

'Believe me, you are the biggest conundrum in the world. No one knows what kick-started you, Kevin. Scientists have tried to create life from the same inanimate particles that you're made of. Recently they used homemade DNA to create a form of synthetic life that makes you look like a dynamo. No one has managed to create anything as amazing as you. It just can't be done.'

'Huh. I'll be damned.'

I tell him about space rockets and IVF treatments, about satellites and the internet and electricity, but when he asks how old the universe is or how big or how Life began, I have to admit ignorance. Not just personal ignorance, but universal. Then I tell him that in the time we've been talking, before he's even blown out his birthday candle, his fellow Kevins have increased by 2,118. I don't tell him every Kevin alive today will die soon, quickly decompose and reappear in some other form. I remind him that he is my grandfather times a zillion. My direct ancestor. (And yours.)

'So, if I made you, how come you don't look like me?' he asks.

I tell him a bit about evolution, but emphasize we're still figuring that one out. That every year an earlier earliest hominid is discovered, and we have to adjust our own timeline again. And sometimes newly discovered bones make us adjust our understanding of what kind of animal we are in fundamental ways. Predominantly nomadic? Social? Bloodthirsty? Altruistic? There's a bear bone with holes in it sitting in a Slovenian museum which might have been a Neanderthal's flute. Or it might have been made by a Neanderthal gnawing to get every last drop of bone juice out.

'It's a constant identity crisis,' I confide.

Pause.

'Can you please stop breathing on me?'

'Sorry. Bad breath?'

'No. I just don't like carbon dioxide very much. Gives me heartburn.'

~~~

So here we are in the warmish un-salty sea: you, me and Kevin. It will be 2.3 billion years before any other life form appears. It's a very slow start to Life and would make a terrible novel.

'But what do you think, Kevin? You single-celled miracle, you.'

I hear a mumble, so I lie down in the water next to him and really concentrate.

'Did you say something?'

'I said, I'm tired now,' he says, and sounds tired.

'Oh. Don't worry. You're just dying.'

'What's dying?'

'The most popular group activity on the planet. And you're the first to do it.'

'So, I'm like a…trend-setter?'

'Correct.'

'Why me? I didn't ask to be born,' he says with a kind of whine.

'I know. No one does.'

'That is the weirdest thing I ever heard of.'

He sighs again, and I sigh too to keep him company even though I'm not sure why. I don't care how weird Life is, I like being alive. I love it.

'Look here, Kevin. Why not just enjoy the party while you still can? From here on in, it'll be bananas. Dinosaurs and ladybirds. East African zebras. Brazilian football players. You're the star of the show. Take a bow!'

'I'm cold. It's getting dark.'

'I'm here, Kevin. Don't be afraid.'

My entire body is wet, but not cold. It would be pleasant if I'd remembered to take my shoes off. Silence. More silence.

'Kevin,' I whisper. '*Say something*.'

'Stop. Talking.'

But now I can't shut up. All I want to do is scoop him up and take him home with me. He's become dear.

'Don't be like that, Kevin. It's not so bad. You may be small, but in the big picture your life is not a small thing.'

No answer, because whatever animated the cell of Kevin has departed as mysteriously as it arrived. His atoms, with their nucleuses, protons and electrons, are still here. His DNA with his unique genes is still curled up inside his nucleus, as well as inside the other Kevins. None of the particles that made up Kevin were ever alive, yet put altogether, for a while, they were.

I stand on the surface of the sea and shake some water off like a dog.

'Rest in peace, Kevin number one.' I say. 'Let us give thanks.'

'Who the hell are you?' comes a multitude of squeaky little voices speaking in unison. They sound a bit like the aliens in Toy Story. Friendly but weirdly monotonic and nasal.

'I'm one of your distant descendants.'

'We made you?'

'Well, not all of you. Only one of you is my direct ancestor, and I'm not sure which. Let's say it's you,' I say, pointing to a single cell clinging on to an identical looking one.

'Me?' says the unchosen one.

'No. I'm pointing at the guy next to you.'

'Yippee!' comes a squeaky voice.

'Aw,' complains his companion.

'But we're all related. We're all the product of the first Kevin.

Who died.' I don't say *for us,* though the Jesus-at-Cavalry echo does briefly occur to me. Apparently you don't get to create Life on earth without giving up life as an individual.

'Ah. *Kevin,*' they all say in a hushed reverent way.

Then a tiny voice I don't recognise whispers: 'How do you know?'

'Because I'm here. And I'm me.'

Pause.

'Aren't you being a bit dim? If you weren't you, weren't here… well. You wouldn't know that, would you? How could you know you didn't exist? Awareness requires existence.'

Is anything more annoying than a lesser being making you feel stupid?

'Okay. But still,' I say, blushing and hearing how childish I sound.

Then the flotillas all shiver in unison, like a giant sigh.

'We miss Kevin!' sing a thousand tiny voices, as melodramatic as an opera. 'He's gone, gone, gone!'

I don't contradict them, even though I know it's not strictly true. I'm tired now, and need my kitchen.

~~~

And then I'm back home. I put the kettle on and open a window, because since I've been gone the sun has come out. There's the world! The snow that fell last night is clinging to each fence post and twig and shrub, and they are all exploding with light. In one form or another every Kevin that ever lived is still here on Earth, I remind myself. There's definitely some former Kevins in my garden. There's some inside you and me. All living things die, but there's no escape from existence.

Polly

Two billion years ago
Still the Ocean

No genitals of any kind yet, but nevertheless: sex is here! After the first ice age, and after a billion and half years of hermaphrodite goings on, of Eves being pulled out of the ribs of Adams, *beings are bonking*. It's got its downside, of course. Ask Kevin's 194,962,272th granddaughter Polly on a Saturday night in the sea off (what will be) Goa. It takes time and energy, and as a prokaryote, Polly is short on both. She's dinky and she's exhausted. Today is her first birthday—not literally a year old, but the prokaryote equivalent.

'You checking me out?'

'Yeah, actually. You mind?' I say.

'Course not,' she says cheekily, and does a little pirouette. 'What do I look like?'

She looks like a minuscule hairy penis with a mouse tail. Not that different from Kevin—still no nucleus, but a slightly more substantial membrane. In fact, Polly is really just an upgraded Kevin. Life on earth is progressing in incredibly slow motion. It's so slow, it looks static if you don't look closely.

But Polly *is* an improvement on Kevin because of sex.

'You look very pretty.'

'Really? You don't think my flagella is too long?'

I don't know what a flagella is, but I assure her it is not too long. It is just the right length.

Look at her.

Okay. Not a good look if you're broody. And I agree—Polly is not the right name for her. For one thing, she's a prokaryote, not a perky pony-tailed Polly. But what can I do? It's not up to me, the naming of ancient ancestors. Maybe Life is programmed to give them the current version of their true name, and like Google Translate, it sometimes gets it wrong. Mostly my ancestors have generic names from twenty-first century Britain and America, meaning (I suppose) that their real names are the generic names of their own era and place. Personally, I reckon Polly doesn't have a name at all. Or if she does, it's something that involves bubbles not words, because she has no mouth, no vocal chords or ears.

Pity poor Polly.

Flitting here and there, swishing her hairs and looking for love. I don't see any takers. If she doesn't hook up with someone tonight, that's it. End of Polly DNA—and the end of the prospect of me. She often wishes she didn't need anyone else. Some of her family are still doing it the old school way, just splitting themselves despite the obvious disadvantages genetically. Some others manage to keep going with transduction, which involves harnessing the environment. It's all part of the ongoing experiment Life is making.

I bet Polly tells herself: *Life must have been so simple before sex*. Just endless quiet replication, no having to consider what colour someone else wants to paint the tidepool. The good news is that she doesn't need to find a boy, because she's not a girl. Gender will come along in another million or so years and make reproduction much more complicated. For now, sex is strictly grab who you can and hope for the best. Not a time for high standards. A time of promiscuity, of sharing your genes and not being fussy. Sexual reproduction—or cellular conjugation, to be scientific—is hard work and Polly would like to know:

'What exactly is the upside to sex?'

I tell her sex gives her a better chance of surviving. Well, not her exactly—her species, close cousins of the eucaryote.

Sexual reproduction could be seen as the first altruistic act, because right now Polly is obviously not getting much out of it herself. Though she's a second generation product of sex, and therefore healthier than her grandparents. Her gene pool has been infused with a variety of characteristics which wouldn't have happened from mere Kevin-like splitting. Maybe 89% of her ancestors died in their first hours because of a vulnerability to infection. With every generation, that weakness would have been reinforced, until almost all of Polly's family were in danger of extinction. Then a few began to duplicate by sharing DNA as well as the usual self-sufficient method of splitting, and voila! The new Pollys survived longer.

'Hence, sex,' I tell her.

'Okay. But is it really worth it?' she wants to know. 'I mean, what's so intrinsically wonderful about us as a species, that I have to make this sacrifice? It's my birthday, for Life's sake.'

My guess is she'd rather be on the sofa watching the primordial equivalent of Big Bang Theory reruns. I sidestep her argument—which has a certain amount of logic—and distract her instead. A strategy I developed to survive motherhood. Also my two marriages. *That's a nice shirt/haircut/pair of shoes. Would you like some lovely mince and tatties for dinner?* (A meal I despise but learned to cook to imperfection.)

'You're absolutely right, Polly. And look, I've got some ice cream for you.'

'Ice cream. You've brought me ice cream.'

Who would have thought a hairy penis-shaped thing could sound sardonic?

'Okay. No mouth,' I say. 'I get it. But it's the thought that counts, right?'

Polly looks at me with even more contempt.

'Who told you that? The thought never counts. Not here. Not ever. Which reminds me. Can you please stop telling people who I am, as if you know?'

'Yikes.' I'm blushing because my imposter syndrome is flaring up.

'I'm not just a projection from your ethnocentric imagination. I'm not unhappy or happy in the particular ways you describe. I'm neither. I'm just trying to get through my days.'

'I'm so sorry,' I say.

She sighs and looks away. Geez. Is she sulking? Some days I feel I'm reliving the worst days of raising adolescents, even though they are now thirty, thirty-four, thirty-eight and forty. It's a little nightmarish, to be honest. I hate being hated. It's tedious. I remember how much my children's contempt hurt at first—who wouldn't be wounded to have the being they'd poured love into suddenly turn on them? But it prepared me for subsequent contempt, like my grandchildren's and now Polly's—so not a wasted experience. Maybe all parents grow an extra layer of skin that forbids *I-hate-you* entry? Those *fuck yous* just slide right off me now. After three seconds of stinging, I remember they love me no matter how much they currently hate me. How do I know? Because I was that kind of kid too. (Sorry Mom and Dad.) Weirdly, these days when I look back on my children's contemptuous years, I almost feel nostalgic. I even smile sometimes, because we were all so melodramatic. Like the time I took my sixteen-year-old daughter to Glasgow to investigate cookery college and dorms, since she'd jettisoned school. I'd imagined us growing close at last, sharing a laugh, shopping for shoes, eating in some atmospheric Italian restaurant where'd she begin to appreciate what a cool mum she had. The reality was that she wouldn't even walk down Buchanan Street with me, instead choosing to stride parallel to me on the other side. It was a pedestrian street and a river of people poured both ways—I kept taking sneaky glances at her, worried I'd lose her, but also proud enough to hope she didn't think I cared enough to keep checking she was still there. She never once looked to see if I was still there—not that I noticed anyway. Her contempt was professional standard. Now she lives behind me and I giggle silently when I see her teens giving her grief. Believe me, every phase of life has unexpected sources of joy.

~~~

The earth's first ice age was half a billion years ago and almost wiped the slate clean—not that there was much on the slate, only microbes in the sea. Some of those Kevins survived—and still do, perhaps they're indestructible—and after a while of being other things, evolved into Pollys. Life is irrepressible, and new species are evolving from older species every day. Bizarre and gorgeous creatures we'll never get a glimpse of, because they'll mostly bite the bullet when the next catastrophic event happens a few dozen years from now. The success of micro-organisms means there's an increasing glut of oxygen—which is toxic for most of life here. Luckily for me, not the kiss of death for Polly. Goes to show how much we can adapt in order to extend our life span. In fact, I'm beginning to think there is nothing we won't do to stay alive. Maybe the species that died off were simply not ruthless enough in their adaptation. Maybe their cells didn't learn to repair themselves efficiently. Or maybe their will to live was secondary to their will to…for instance, be creatures of habit. Altering one's self is stressful, but the alternative is extinction.

Look around. Life forms are not the only things that continually change. The sea bed, the water above it, the air above that—their molecules all wriggling around, eroding, re-forming, and in air's and water's cases, even dancing. Huge rocks at the bottom of the sea, looking so solid—give them a few millennia and poof! Grains of sand. Does Earth have a life span too? It has a beginning and in my experience beginnings are preludes to endings, once the middle bit is over. Since I started thinking about evolution, I've been thinking about my relationship with the world. For at least four decades, it was like...well, like the way I felt about that comfy old armchair, the one I've had forever. I was aware of it, but only peripherally. A dull backdrop to my exciting life, but now—my God, that chair is amazing! I'll never get rid of it. And so it is with the world. I'm not young. I'm not even middle-aged anymore. It's impossible not to feel a rush of affection every morning when I open the door and stand in the garden. *Wow. There you are again*, I think to myself. *Hello world!* My familiar view with the sun breaking through the mist over the Firth, or already high in an August sky. And ten minutes or so later, deliciously bitter coffee on my tongue, down my throat, making my heart pump that much harder for a while. I'd been told the senses get duller as one gets older, but I've not found that to be the case. My morning coffee just gets better and better, as does my evening glass of red wine. Daily sensual indulgences. The average life span for a woman in the UK right now is just under eighty-three years. When I was young, I thought eighty-three years was a long time. Imagine that.

~~~

Homo sapiens have only been around for roughly 200,000 years. We're infants. The average lifespan of a species is four million years, and 99.9% of the species who've existed are already extinct. They evolved until their adaptations failed to keep them alive. Or, of course, until a catastrophic event happened, or another species came along and didn't want to share resources. Much random luck is involved. It seems inevitable that we'll cease too—and theoretically, some days, this feels acceptable. Extinction is natural and we must accept our fate humbly. We are momentary blips, our atoms destined to form other things when we're gone. Some may become part of intelligent life forms who'll study our bones and make wildly inaccurate and accurate assumptions. *A dominant species of primate, whose success and downfall were both due to excessive territorialism.* They'll be aware they partially comprise the things they're studying—atoms from you and me—and the irony might tickle them. It does me, anyway—because that is what evolutionists mainly do.

Okay, I say to myself. *So be it.* Homo sapiens will have had their day. But when I picture individuals I know—perhaps a particular grandchild with her innocent expectation of a life, her excitement when she talks about getting a new bike—imagining extinction is unbearable.

Oh, the very oddness of *us*. The current dominant species on the planet, with our high foreheads, our large brains, our self-obsession and endless self-analysing. The high value we place on being alive. Has any other species been aware of their existence and of their decline the way we are? Studied it, recorded it, quantified it? It's hard not to feel superior and weep. As if because we will watch our own exit, it will be a much more tragic end.

~~~

But we are not here to think about the end of Homo sapiens. It's Polly's birthday! A time to celebrate survival. To make a long story short, Polly needs a bloke. So to speak.

'Good luck, Polly. If it makes you feel any better, sex will always be a mixed bag.'

I'm thinking of my unreciprocated crushes over the years, and of that weird menopausal spell of wantonness. So much insanity and shame with sex. My God, it's got a lot to answer for.

'Thanks. That doesn't help.'

'Just close your eyes and think of something nice.'

'Like what?'

'I don't know. What's your favourite thing?'

Pause. I try to imagine what her options might be. Kevin-counting? Doing wheelies underwater?

'I like love stories. Will you write a romance with me in it?'

'Oh, please,' I say and tsk in as disapproving a manner as I can. Not only is her request bizarre, it's sentimental. 'You are a procaryote, my dear. You can't possibly yearn for romance. I must be projecting. Just ignore me and be your sweet single-cell self.'

'See that sunset over there? Let me walk into it with my soulmate,' she pleads, as if I haven't said anything. As if she has legs to walk on. 'Please, please, please!'

'Listen, Polly. What you're doing right now *is* pretty romantic, in a bleak original way—which is actually much more romantic than strolling off into any sunset. There's pathos in the unflattering unfettered truth. You're trapped by your nature and biology to bonk anything you can get your hands on right now. It might seem a heart-breaking scenario, but it's got a certain integrity. You're trying your best with limited options, and what's more heart-warming than that?'

'You're just telling me a story to make me feel better.'

'Well, feel better then!'

Then I lower myself until I am right next to her, and whisper: 'Listen, that's how I start every day—tell myself a quick story. A convincing and hopeful narrative of my life so far, just to make sense of the world. Of my place in it. Why not believe the version that makes you feel good?'

'Can you please back off a bit? I worry you're going to swallow me. Breathe me in or something grosser.'

'Sorry.' She reminds me of Kevin, which shouldn't be surprising.

'And I still want the sunset story,' she whines. 'It would make me feel much better than the version where I have no standards and end up bonking my fat cousin with bifocals.'

'Jesus. Well, alright. A sunset it is. A lurid sunset with you tail in tail with Mr Tall Dark and Handsome.'

I smile to show I'm not offended, even though I kind of am. Not only has an ancestor of mine got Mills-and-Boone taste, her fat cousin with bifocals is probably my grandfather times a zillion.

'Yes, yes, yes!'

'I prefer not to judge, Polly. To respect each of my forebearers equally and with...'

'Shush,' she says suddenly, and I notice her hairs are quivering. 'Need to be alone for a while. Do you mind?'

'No. Not at all.' But I can't resist adding: 'You've spotted someone nice after all?'

'If you call possessing genetic material *nice*.'

'I do, actually.' Especially if it ends up in me being me—which is so nearly NOT about to happen again.

And that's it. Polly disappears and I'm alone.

'Be good to each other,' I whisper in the general direction she's gone. And then in a louder voice I say: 'Hey. Thanks!'

From a distance I hear some coy giggling.

William

555 million years ago
Offshore what will be Morocco

The Kevins and Pollys have been twiddling their thumbs. Here we are, 250,000,000 whole years later and we've finally got William. What's so special about him? He's got more than one cell, for one thing—quite a few more cells, but his main claim to fame is being our earliest animal ancestor. Hold your breath, because he's on the sea floor, and don't worry about a wet suit because the sea is still warm. Okay, open your eyes now. Yup! William is a worm. Every single animal, including us, owes their existence directly to a worm. But quite an interesting worm, as worms go. Something going on with his colour that's quite enticing, don't you think? Greyish, then brownish. No siree, nothing boring about William.

There's a jellyfish kind of thing nearby who's not impressed. She claims *she's* the origin of all animal life, not William, and who's to say she's not? And there's a thousand or so sponges who've been trying to stake that claim as well. They've been organising themselves into little protest groups, carrying signs like: *We're Number One! Suck it Up!*

The contenders for first animal all have this in common: They're small, they have soft bodies, they're genetically complex, and they have a talent for keeping very still. Sometimes they move, but so slowly you'd hardly notice. It might be something they learned to do during the recent ice age when they had lots of time to kill and much reason to conserve energy. It seems ice ages are not entirely bad things, in the same way forest fires generate new growth. The first ice age resulted in cells no longer only splitting asexually. The second ice age resulted in life forms that were more complex, with more than one cell.

'You mean I don't look like a hairy penis or a bit of jelly?'

'Gosh, you've been listening?'

'Naturally. You're my first visitor in….ever.'

'Well, you're correct, William. You actually look as worm-like as a…a worm.'

'Which is an ugly thing to be, I suppose.'

'No, it's a handsome thing to be. You're really handsome, William.'

'If you say so,' he says, sighing.

'I've come to wish you happy birthday.'

'That makes me happy.' He says this in an unhappy voice, so it's kind of funny. But I don't laugh. I'm not sure what to make of him. I wasn't prepared for a worm to be melancholy. Though he's an animal, so why shouldn't he be as moody as any of us? He has a proper nervous system. He senses things and responds to them. He eats passing vegetation, extracts oxygen from the water, reproduces himself with the help of someone called Wilma, and sometime later he will die. Like the Kevins and Pollys, worms like William are still around now and haven't evolved much—mainly because there's no need. The worm might be the most robust product of evolution so far. If it isn't broken, why fix it? Look at him. William has a brain, he feels pain, and he knows how to endure. He does not, as far as I know, fart.

'What's a fart?' asks William.

'Exactly,' I say.

He sighs again and then slumps, momentarily reminding me of a boy I once dated. He was always stoned and had a limited repertoire of facial expressions. 'Wanna lay on the bed and mess around?' I'd say. 'Oh, man,' he'd drawl, like just speaking was tedious enough. 'Alright.'

~~~

What happened to make cells so smart? It's hard to imagine them specialising without anyone telling them what to do, without the ability to think or the capacity to take a decision—or indeed, an action. They're so tiny, and made up of atoms which are inanimate particles. But think of a human sperm and egg. They meet in a dark wet place, just two microscopic scraps without a thought in their heads, and wham! Add a steady flow of nutrients from the placenta, and nine months later an entire human being has been constructed—complete with eyelashes and toe nails, as well as proclivities like an ability to play piano by ear, and weaknesses like a susceptibility to asthma. This happens about 250 times a minute in the current world. It's so commonplace, we forget how incredible it is. How the hell do those cells know what to do? At a slight remove, how the hell do individuals in the same species know to specialise in lifelong passions as diverse as lepidopterology and dairy farming and palaeontology? Does every expression of Life follow the same pattern? Survive, reproduce, fulfil a function benefiting the species - preferably not an over-subscribed one. I wonder if any groups of Homo sapiens failed because everyone wanted to be the equivalent of a lawyer or a doctor? Who would grow the food, who would nurture the babies and collect the rubbish?

But on an embryonic level, which is far easier to study, scientists say—as if this makes it less miraculous—cells know what to specialise in for the same reason William knows what to do now. Their destinies are implanted in the DNA curled inside them like coded secret messages not to be opened until fertilisation has taken place. *Wake up. Your mission is to become a fragment of liver. Focus. No more treading water.* This applies to all cells except, of course, the very useful pluripotent cells. They lack any kind of instructions whatsoever and become whatever the current chemical signal deems they should be. Like unformed fourteen-year-olds responding to the current trend in teen slang or morality or trainers. Bottom line? William didn't plan on being a worm, any more than you had a say in being a human being. No cells consulted him beforehand, and no atomic precursor of cells consulted the source of cells, asking to be a cell. Kevin nailed it over a billion years ago—no one chooses to be born. It's something that happens to us (though sadly some people wish it hadn't.)

~~~

'Hey ho,' says William when I tell him all this. 'Big deal.'

He wants to know if I have anything for him to eat.

'It is, after all, my birthday.' His tone is so doleful, I think of Eeyore.

'I've got this cake for you,' I tell him, and pull it out from my pocket.

'Call that cake,' he says, not asks. But he has a point. The cake is soggy because I'm underwater.

'Sorry,' I say, even though I feel he's impolite. Would a tiny *thankyou* be too much to ask? So far, the early products of evolution are the last to be impressed by the process. Or even by my presence. I expect surprise and awe, to be honest. Secretly, I even expect a bit of love. I sing him one verse of Happy Birthday. Not surprisingly my voice has an aquatic warble, but I carry on anyway. Then I plant a kiss on him—this sliver of unsung slime in the murky depths who may or may not be the first animal on earth. He's one today—or the worm equivalent—and that's worth celebrating no matter what. Any survival at this stage of the game is a marvel. A whole year! Naturally, he's disgusting to kiss.

'Happy birthday, William. And happy birthday to all your cells, too. Clever, *clever* cells, to end up making wonderous you!' I discreetly wipe the slime off my mouth.

'You're so patronising.'

'Sorry.'

Pause. How many times do I have to say sorry?

'It's alright.' He stretches this out, so it sounds like awww-wlllllrrrrriiiiitttte. Again, I think of my stoned boyfriend who didn't even notice I'd dumped him for six months.

Then William sighs, letting a small bubble escape. I'm not sure, but he might be a little depressed. I don't tell him he's going to die one day. That would just be mean. Thinking about him dying doesn't make me sad. He's already reproduced, so my ancestor's lineage is intact. Ce la vie, William.

Betty

400 million years ago
Near what will be Tahiti

Welcome to the Age of Fishes. Thanks to the recent surge in oxygen, life here on Earth is going a bit crazy. And by earth, I mean the sea where everyone is living. And by everyone, I mainly mean the euteleostomi, or bony vertebrates. Fish with bones. More than 90% of sea residents belong to the euteleostomi clade. And by clade, I mean a group in which each individual can claim a single ancestor in common. Yes, your grandparents times a million are here in the sea too. You might even be Betty's direct descendant, like me. So next time you feel at home in the sea, remind yourself it was your family's home for almost 400,000,000,000 years. It's why you hiccup, in fact. An evolutionary remnant from when our gills gulped air from water. It was only 375 million years ago that our ancestors left the sea. And remember—Homo sapiens have only been around for two million years, give or take a century. If you don't happen to feel at home in the sea, well, home is not always a happy place for everyone. Maybe your DNA is saying—*No, no, no! Get the hell out of here! No regressing!*

~~~

There is no life on land yet, and very little land anyway. In about 25 million years, one of Betty's descendants called Margaret is going to take it into her rebel head to use her leg-like fins and crawl out of the sea. But for now, our clade is happy swimming around this neck of the woods—which since William's day has become a bit salty. Betty is like a medium sized salmon and has a greenish tinge to her scales. She's considered the ancestor not just of us but of all amphibians, reptiles, birds and mammals. To date she's the most sophisticated creature ever. Quite pretty, in a way. Lovely large eyes and a long soft textured body—one might even describe it as paisley-patterned. The pattern ends neatly two inches from her face, which gives her a tailored look. She has tiny fins at the front which in many millions of years will become arms. Her two fins at the back are larger and will one day become legs. She already has a nasal cavity like ours, and three lobes to her brain. (Only one more to come!) Her mouth is wide, with sharp pointy teeth. Quite scary if she's hungry and you're a little fish, but today she's smiling like Mona Lisa.

'So, Betty. What's up?'

'Who are you? What do you want?' Slightly aggrieved tone, like *Can't you see I'm busy?*

I had a boss like her once. I used to tiptoe around her. Not because she made me afraid, but because I'm a slow thinker and knew I'd never think of the right thing to say back to her until hours later. Awareness of one's weaknesses leads to strategies like avoidance. But boy, it was exhausting some days. It would be so much better if I was a fast thinker, and also a person who didn't mind not being instantly liked by everyone. The burden of my own nature seems too heavy to bear. Oh well! What a self-pitying wimp I am.

'I'm one of your descendants,' I say self-consciously, having all these negative thoughts.

'My....distance?'

'I'm Cynthia. Nice to meet you. I don't really want anything.' I smile a little. Her ignorance of the word *descendant* has cheered me immensely. Another example of how pathetic I am.

Suddenly she is no longer like my bossy boss of yore. I hold her left front fin briefly in my fingertips—it's rough and ribbed. I feel it rapidly twitch and drop it.

'I don't like to be touched.'

'No problem.' I'm relieved, actually. 'I just wanted to say happy birthday.'

'Huh. Huh. Huh.' She swims in a small circle as she *huhs*, using the two fins on her left side to do this. Think of turning a row boat using one paddle.

'You're having a nice day, right? You seem happy.'

'Yeah, well I am, as it happens. I just laid my eggs over there, and let me tell you—it's a great feeling, getting the eggs out.'

Now it's my turn to say: 'Huh.'

'Like having the biggest bowel movement you've had all year.'

'Gosh. That does sound good.' Thinking: But aren't bowel movements all fish ever have? Do fish pee? (Yes, I know they must, but have you seen a fish pee?)

'And look, here he comes. Shush, don't distract him.'

I hold still, hold my breath—which is easy, as I don't generally breathe under water anyway. I stare as my other direct ancestor hovers over the eggs, then squirts a cloud of sperm over them.

'Don't move!' hisses Betty. To my knowledge, this is the first time an ancestor has hissed at me. I shouldn't be surprised, each one being unique. I could never mistake Betty for Polly, or William for Kevin.

'I'm not!' I hiss back. What a nag.

'Were so.'

I open my mouth to argue, then think—no point. Now Betty is reminding me of my sister, with whom all arguments are disasters for me. Probably because she's usually right. When the sperm cloud disperses, mostly into the current but enough onto the eggs, Betty turns to me and smiles…post coitally. There is no other word for it. Then she coughs, and it reminds me of a smoker's cough. Maybe pretend-smoking is something she always does at this juncture.

'Want some water?' I ask.

'Are you insane?'

This is a question I'm asked a lot. I don't take it personally. I let some silence pass. Twiddle my thumbs and watch the shadows passing over the surface of the sea above us. When I finally turn around, Betty looks asleep on the floor of the ocean. Not dead, because her fins give little flicks now and then to keep her from drifting, but otherwise motionless.

'Are you still here?' she asks, eyes still closed. 'What do you want?'

'I want to say happy birthday Betty. That's all.' I don't remind her that I've already said this. I don't want to imply she's stupid or has early onset dementia.

Silence. Then:

'Do I need to know what happy birthday means? I'm not in the mood for thinking hard right now.'

Pause, while I consider the benefits of comprehension.

'Nah. Don't worry about it. It's a good thing, that's all you need to know. I'm glad you exist.'

'Well. Me too, obviously. Glad I exist, I mean. I don't give a shit about you.'

I want to tell her it's exciting to meet her, because she didn't seem real to me before and now she does. A fraction of a fraction of her DNA is inside me right now. And I do happen to love floating in a warm sea. I never want to get out. But she looks so peaceful with her lovely eyes closed, I keep my mouth shut. Instead I dog paddle over to the eggs and salute them silently. *Go eggs!* I think.

Margaret

375 million years ago
Near what will be Greece

We're three minutes before the appearance of Homo sapiens, if Life on earth is crammed into sixty minutes, and all I can say is: *Thank God for brave ancestors.* The sea was the only habitable place for three billion, six hundred and fifty million years, give or take a millennium. That's all of Earth's existence, minus a billion years, of things not living on land. Which made sense of Margaret's choice of abode, until last month when she crawled out of the ocean on to the only land mass currently on the planet—Pangaea. Which is not the first solo continent. Already, over the last 4.5 billion years, there's been at least two other super continents. These were broken up and reformed because Earth's mantle *will not* settle down. Like a kid with ants in her pants. (Or me, when I was younger and moving constantly because I'd read *On the Road* and wanted to be a girl version of Jack Kerouac.) The mantle comprises 84% of the planet's volume, so when she decides to go walkabout, everyone knows about it. And yes, the mantle still has wobbly days now, days when she wants to strut or do the shimmy and to hell with everyone else, but we mostly don't notice. From our time perspective, she's a zillion times slower than molasses poured in January.

~~~

We're here to visit Margaret. The first ancestor to crawl out of her ocean home.

All alone. With no idea she'd survive the day. No idea, even, of what lay in wait. It might have been hundred metre boogey octopi, magicked to shore and laughing maniacally. But Margaret kept right on coming through the shallows, half swimming, half dragging herself. I imagine she paused half way, sunk down in the mud and absorbed the strangeness of it. The gradual tightening of her skin as the sun dried it. The blinding glare of light on the surface of water, instead of being diffused by it. The sight of green vegetation over there in that…that non-water place. Then she took the monumental decision to take air in through her mouth instead of her gills. What inspired her? It would be like me deciding to put food in my ears instead of my mouth. It was a huge risk, because who was to say there'd be the right kind of oxygen outside the sea? Indeed, when considered this way, it's pretty audacious for human babies to confidently exit their briny world after nine months. Are their lung cells psychic? Do they sense they'd find the right kind of oxygen away from the blood pumping in from mum's placenta? How do babies know these things? Maybe this is why some require forceps or caesareans. They're simply being cautious. Risktakers, all the rest of us.

But Margaret's luck was in. She inhaled again, then exhaled her own mix of carbon dioxide, which was immediately absorbed by the plants which had crept out of the sea a hundred million years earlier—flora being far more advanced than fauna. Just like her, they'd found themselves stretching out towards the sun, creeping on to the land, eventually becoming trees and flowers and shrubs giving off oxygen. All of which made dry land habitable for Margaret's scary adventure. *Thank you plants*, Margaret might have said that first day if she'd known. If she'd had both awareness and manners. *Thanks for getting the place ready for me. It's nice here. Very cosy.*

~~~

Anyway, that was eight months ago when she was four months

old, or the equivalent of. Now she's living here, on land. Still near the sea, and sometimes she slips back in the water for a moment or two—but her permanent address has a land postcode. The new kid in town, although not entirely alone anymore. The urge to head landward is rising everywhere, which means the competition for territory has begun. Yesterday she was almost eaten by something with two heads and a blue tail, another recent sea migrant—but she's surprisingly quick. How narrowly I almost miss existence, over and over.

Look at her.

The Canadians who find her in the twentieth century will call her Tiktaalik, which kind of suits her, but it's not her name. Margaret, like me and possibly you, is descended from a fish called Betty (you've met her), but doesn't resemble her much. Margaret looks more like a Dr Seuss character. Four flapper legs, a mermaid's tail, a crocodile head, purple and blue skin, and…is that a smile? Hard to tell. Might just be the permanent shape of her mouth, which stretches all the way around her face. She's about two feet long and a faster mover than you'd think, given the size of her legs-which-used-to-be-fins. Knowing she's one of my family, of course I see her as beautiful. *You are beautiful!* my mother says in my head. *Thank you*, I reply. *I get that from my mother.*

~~~

Disclosure: Not all my direct ancestors are cute babies, and a fair number of them grow into adults with horrible personalities. Edwardian Edwin is one of the worst, and his crimes—while not worthy of prison—are still too vile to mention. The way Bertie O'Shea likes to touch his wife's little sisters bothers me too, and Doris the cat tormentor from Dorset is just despicable! I'm not writing about them for the same reason I wouldn't invite them to my dinner party. I don't like them! I'm mentioning them now in case you're thinking *Jeez, she's making all my relatives look like dolts and numbskulls.* Believe me, I have a large smattering of dolts and numbskulls in my family too. If, on the other hand, you've been thinking: *Golly, I'm sure glad I'm not in her family! What a bunch of losers!* then ignore this paragraph. Possibly close the book now. You can leave it on a park bench or bus seat, I don't mind. I truly understand.

~~~

So Margaret is not alone anymore, which is good. Being the only one of your kind might be the loneliest thing there is to be, which makes me think of my early ex pat days when I was a barmaid in a tiny country hotel near Fraserburgh. I thought I was happy and assimilated until an American man walked into the bar one day. I gravitated towards him despite awareness that if I'd seen him in San Francisco, I'd have given him a wide swerve. He was utterly not my kind of guy, yet just hearing his accent pulled on me in some visceral tribal way. I was quietly shocked at this apparent evidence of my homesickness. Then I think of Kevin the original microbe, who was also the only one of his kind. I wonder if Loneliness with a capital L was what motivated him to split into two that day. Maybe Life itself was lonely, after getting over the excitement of being alive. I mean, who would you celebrate with? Have a moan with? Maybe Margaret's courage encouraged others—not just her own species, but other sea folk of all shapes and sizes. Or maybe everyone just had the same crazy idea about the same time. Hard to believe it's only been a hundred million years since the second ice age obliterated everything. It's early morning and I can already tell it's going to be a hot day. *Listen.* The air is full of whirring sounds from insect-like things, and swishing sounds from I-don't-know-what moving through tall grasses. Margaret, however, is silent. She's watching a dragonfly-like insect which is about twenty-eight inches across.

I'm afraid to distract her, she's so intent. I wonder if her mouth is watering.

'Margaret,' I whisper.

Nothing. She doesn't even turn her head. Then, in a blur of motion, the dragonfly-like thing is in her mouth and she's munching away with such evident pleasure, for a moment my own mouth waters.

'Wow,' I say. 'Does that taste good?'

She looks up at me, but says nothing. Stops chewing.

'It's okay. I'm not going to hurt you,' I say. I crouch down to be near her. 'I just want to say happy birthday. And thank you.'

'What for?' She tilts her head as she says this, and loudly swallows. Gulp!

'If you hadn't crawled out of the sea, I'd not be me.'

Silence.

'Are you a poet?' she asks.

'No,' I reply, slightly insulted.

'Sorry.'

'I was just telling you the truth. If you'd not crawled out of the sea, this book wouldn't exist because I'd not exist. I wouldn't not just be me, I'd not be.'

'There you go again. If you don't want to be called a poet, stop talking in rhymes.' She tsks.

I decide to ignore this. She's probably nervous, hence the attitude.

'Remember those eggs you laid in the shallow bit of a pond last week? Well, you're my grandmother times a zillion,' I continue, undaunted.

'Wait. I'm pregnant?'

'No, not exactly. You don't have the apparatus for pregnancy. No one does yet. But a cousin of yours was watching you lay those eggs, and later, well, when you weren't looking, he crawled over and...'

'Please, please! Too much information.'

'Yeah, okay. It is pretty weird, though, isn't it? You lead directly to me.'

'That's nuts.' She makes a nasal bleat which I interpret as a contemptuous laugh.

'Not really, when you think how long it will take. You will evolve into other life forms, and eventually they won't look the least like you. A few of them will survive the next ice age and the catastrophe after that too, when the world goes dark for a year.'

Pause. I wonder if I've disclosed too much. I don't want to plant any seeds that might stem the flow of evolution—for obvious reasons. I don't want to be un-born.

'You're so funny looking. I mean, what's wrong with you?'

'Nothing's wrong with me, Margaret,' I say, relieved she hasn't really been listening. 'But I agree, I am a little funny looking. Both my parents were attractive, so I don't know what happened there. There was a red haired milkman who used to...'

'Can you please stop talking?'

'Oh. Of course. All I wanted to really say, is happy birthday. Also, you're pretty gorgeous.'

I notice there's bit of dragonfly-like wing on her chin, but don't say anything. I wait for her to appreciate my appreciation. Inside, I have a glow, despite the insults and indifference. She looks at me blankly, then twitches and says:

'Oh look, there's another one!'

I look where she's looking and see another dragonfly-like thing.

'Do you want a birthday hug? Or birthday bumps?'

'Nah. Got to dash,' she says, slithering towards the dragonfly-like thing.

To be honest, I'm relieved. I would have done it if she wanted, but hugging Margaret would not be a cuddly experience.

Within a second, she's disappeared into the undergrowth. She can't see me anymore but I take my hat off to her anyway, and bow to my ancestor who had the audacity to leave home. Maybe she was storming off from her baby version of a husband after a fight, with him shouting *Don't you walk away when I'm talking to you, damn it*, but even so—Margaret has guts. The unknown called and she answered. Call me an ancestor-worshipper, but who wouldn't worship a creature with that kind of optimism and courage? Kind of reminds me of my immigrant great grandparents, coming from Ireland and Italy to settle in a lawless new country. And in a smaller sense, of my mother and her sister, when as teenagers they left the valley for wild San Francisco. And in a less noble and even smaller sense, it reminds me of me at seventeen. My father ordered me to not befriend the new hippy neighbour down the block. I immediately knocked on that neighbour's door. He was about twenty-five, or maybe thirty-five, broad and pony-tailed. I asked him if he had any rolling papers, which was the only chat up line I had in those days. Within minutes we were stoned, within the hour we were engaged in sex, and within seconds of walking back in the door of my own home, my father threw me out. Did I apologise and beg to stay in the bosom of my safe suburban family? I did not. I was not a remorseful or responsible daughter. I packed a bag with essentials and hit the road. Which was, in retrospect, one the dumbest things I ever did. So maybe I was brave like Margaret, but definitely not as smart.

Beryl

200 million years ago
Bristol when it was somewhere else

It's midwinter and we're in England, and yet...it's hot. That's because England is not England yet, and it's definitely not where it is on the globe today. Not by a long shot. It's roughly parallel with where Tennessee is now. Or northern Africa, hard to tell. We're in the northern part of the subtropics, hence this glorious sun. Oh, and we are not an island yet. There's only one land mass on Earth again. (Fickle old mantle!) It's a lovely mostly fertile continent and we're in the middle of it. There's a huge mountain range through (what will be) Britain, reaching thousands of miles east and west. Those hills over there will one day be called the Mendips. *Lovely place to go walking*, hikers will gush. *Some nice pubs nearby.*

~~~

You think some things in life are permanent and simple. Ha! Nothing on this planet is simple or permanent. Land masses have always moved. The ground is moving right now as you read this, not much slower than your fingernails are growing. The North American continent and Eurasia are moving away from each other at a rate of an inch a year. Think of it like a ballet danced in very slow motion. Graceful, magnificent, sombre. But sometimes there are pivotal events, and those are never like a ballet. Earthquakes, volcanoes, tsunamis, hurricanes, meteors, asteroids—these create physical change quickly and unpredictably. I can't help but think of my volatile children. Toddler tantrums like volcanos erupting, adolescent sulks like fault lines just before a quake. But for the most part, Earth's mantle chugs along in seemingly-motionless motion.

~~~

I've visited before and I know today something exciting is going to happen. That's not the only reason I'm here, of course. It's Beryl's first birthday and she is my grandmother times a helluva lot. Lucky timing for me. Look around—no sea in sight, right? Nothing but land as far as the eye can see. My feet feel solid on the ground. (God, I love it when a day doesn't go according to plan. What wakes you up more than the unexpected? What stretches time more than newness?) Beryl has no idea, of course. Look at her. Darling girl. She's not a mammal or a dinosaur, but the needed evolutionary link to both. Short bow legs like an alligator or a newt. Armoured thug face like a rhino. Tiny ears which flick towards sounds. (I'm jealous! I wish I was one of the 15% of people who can still wiggle their ears.) She has a short thick tail. No fur, just a thick hide with a pattern of tiny dots—aside from her underbelly, which is pale white and looks ticklish. She's a therapsid, and from here on in (again) Homo sapiens are inevitable. Or maybe that's just retrospective arrogance. She's already six feet long, which is big for her age. The other kids tease her a bit. It's a good thing when the male is big for his age, but the longish girls get a hard time. Maybe because in reproduction terms, a smaller female is more likely to escape being eaten. Later, in millions of years when reproduction requires proper sexual intercourse, a smaller female will be more likely to be overpowered by a sperm-producing male, and therefore more attractive. But for now, Beryl is considered less viable than a smaller therapsid simply because smaller equals faster and more easily hidden, and therefore her eggs are more likely to be fertilised.

'What, you think I won't have babies?'

'Not at all, Beryl! You'll have a grand family.'

'Darn tootin I will.'

'But why are you even thinking about this now? You're only one. You must have tons of more important things to think about.'

'Are you nuts? Nothing is more important than my eggs becoming miniature me's.'

'Huh.' She has a point I guess. Though not much of a childhood.

'Beryl, where is your family right now?'

I can hear some chatter over there by that clump of yellow flowered trees, but it's too far away to see if they're therapsids too.

'No idea. I don't need them. I'm one! Why?'

'Oh, no reason. But it might be a good idea to stick close to them today.'

'Because it's my birthday?'

'No. Well, okay. Sure.' Mustn't tell her the real reason.

'Tell me what? Wait. Is this about the continental drift thing again?' She says this like I've been naughty to bring it up. Accusingly.

'You know about it?'

'Christ, it's all anyone talks about. *Bo. Ring.* And they make us feel so guilty about it. Like stop tossing all your bones just willy nilly, you'll break the world apart, and then where will we be?' She says this last phrase in a mocking nasal tone. Oh! She's so disrespectful! (Was I like this in my early teens? *Oh God.*)

'Interesting!' I say, concealing my annoyance. 'But it's not the continents drifting, you know. They used to think that, but these days it's more about plate tectonics. We aren't floating, we're being pushed.'

I think this is fascinating and I look at her expectantly. Silence, then a huge therapsid sigh. If you've never heard a therapsid sigh, think of an asthmatic menopausal woman who's pissed off about something. The air conditioning is broken *again*. They've run out of toilet paper *again*. A deep but also trivial hopelessness.

'Whatever,' Beryl finally says, post-sigh.

I want to explain about the mountain ranges under the sea. About their volcanoes erupting into salt water and lava spreading before solidifying on the sea bed—which is, of course, part of Earth's mantle. About this cooled lava gradually building the sea floor up and pushing the large slabs of rock (which are the tectonic plates) on which our land sits.

'Hey, want to hear something else bizarre?' I haven't given up yet. I *will* excite this cynical kid.

'If I must.' Eyes rolling.

'No one knows for sure—in fact, there's not much anyone knows for sure about evolution—but there's a theory now that the oceans...'

'Ocean,' corrects Beryl in a prissy therapsid tone, her lips curling in disdain.

'That the ocean seeped out of the Earth's mantle. That there might be...' And here I drop my voice dramatically for effect. I whisper: 'Beryl! There might be a huge reservoir at the centre of Earth right now. An ocean without light. Without sky.'

'Without life?'

Oh my God. She's into this now. Look at the scaly hide on her sides—it's fluttering like her heart's beating fast.

'Who knows. Maybe there's a whole world of life in that sea for whom darkness is normal. Maybe it's an advanced civilization and they speculate there's four-eyed purple life forms up here. Maybe they say outer mantle like we say outer space.'

But she's rolling her dinky dark eyes again and yawning. Jesus, her mouth is huge. I can see right the way to her gullet. Not pretty.

'Are you finished now?' she says, when the yawn is complete. I pause breathing because the air stinks of whatever life form she ate an hour ago.

'Well, pretty much. I guess.'

'Because I got to take an enormous dump.'

Are all my ancestors obsessed with bowel movements? I have a brother who talks crudely like this. He's funny because he never tries to be funny, plus he has a face like Alfred E Neuman from Mad magazine. He just states his case with a deadpan voice. Makes me laugh every time. Almost.

'Ah. Right. Well, before you go, let me just say happy birthday again. You are my grandmother times a zillion. Roughly.'

Pause while she licks the insides of her left nostril with her tongue. She eats whatever she's found. Then she looks around with an indifference not often seen on a kid's face.

'Huh. Some days are just weird, is all I have to say to that.'

'That's a fact. Have a nice dump,' I say, and smile.

Not too difficult, despite her lack of appreciation of me. I'm too humiliated to give her the small gift in my pocket. I can't risk any more rejection. But that's okay, I've kept the receipt. I watch her waddle off (therapsids have no grace whatsoever) in the opposite direction of her family. I'm about to shout to her to turn back, when it happens. It's been such a long time quietly coming, but now the change is instant. One minute there is unbroken ground between her and me, the next minute a crack appears and the air is full of snaps, creaks, groans. The sound of the tectonic plate ripping itself in half right in front of me. Somewhere sea water will be rising with a hiss and roar. I can smell it. Brine and the mostly-methane gases from trees and plants and soil being violently disturbed. Not to mention all the terrified life forms whose adrenalin is on full throttle. (Oh yes. Fear has a smell and it isn't sweet. Sour milk crossed with old urine and something else unpleasant.) Everything from the mantle up is being sundered and the west side of Britain is born as coastal. Terrible destruction, but simultaneously magnificent. Oh, more than that. Beryl may be a pain in the butt, but I am definitely going to keep coming back here. Wow. What a show.

I look around for her and there she is, trying to get to her family and crying like the one year old she really is. Wah! I forgive her all.

Ollie and Harriet

66 million years ago
A place that will be Ukraine one day

Much has happened in the 130 million years since Beryl witnessed the birth of Britain as coastal. Another ice age killed off most living creatures, including most of her descendants—but not the one who was my ancestor (whew!). A few millennia after that, the volcanos went berserk for no obvious reason, leading to a long period of darkness. No photosynthesis meant no food. Another zillion life forms we will never know about bit the dust. Earth's slate has been wiped clean four times now. Reminds me of when God was someone the Incas called Viracocha, who lived in Lake Titicaca. He created everything. He made humans by breathing into stone, and whenever they acted too dumb he wiped them all out and blew into some other stones, hoping for an improvement. Luckily, Life—still holding the strands of the DNA ultimately destined for you and me—laughs in the face of things like ice ages and volcanos. *Ha! Think that'll stop me? Think again, punk.*

~~~

Look around you. Even if we don't turn our heads, there are 139 species right in front of us. You won't recognise most of them. Some have three eyes, at least one has one eye, and that green flat thing on the leaf has no eyes and doesn't care. Lots of creatures seem large and therefore alarming. There's a fluorescent pink scorpion-type thing over there in the shade by the rock, see him? He's almost eleven feet long, and could dismantle you in seconds. It's a very dramatic, almost ostentatious phase of

Life. Such vivid colours, such extravagantly sized plants and creatures. *That's not classy*, my mother would say dismissively. I just think of it as a little over the top. And because my species tends to want to 'improve' what it finds, I'd like to tone everything here down a bit. Make it less gaudy, introduce some greys and browns, some humility and solemnity. But suddenly something shifts in my perception and I find it all invigorating. Much better than the tepid magnolia-painted world I live in. Ridiculous pale uninteresting me, and why on earth does anyone want to be classy? I've just arrived at a fabulously trashy party and realise I'm vastly under-dressed. Everywhere I look I see eye-catching oddities, but most visible are the dinosaurs.

Dinosaurs!!!

Some of them I vaguely recognise from reconstructions in museums or films like Jurassic Park, although none of those accurately convey their shininess and stinky-ness and spookiness. Some are about my size or smaller, but mostly they are huge. Dinosaurs are obviously the boss here. What gerbils are to us, mammals are to dinosaurs. Vaguely interesting but vastly inferior beings. Being around dinosaurs is a challenge to my self-confidence, which is never good at the best of times. When I first came to Britain at age seventeen, my American personality hunkered down. Whenever I opened my mouth, a big part of me cringed and tried to disappear. I began to talk so softly people were always asking me to repeat myself. I hated the way I sounded and looked. So gauche. With all my being, I just wanted to be British—a vastly superior species, to my mind then. More intelligent, subtle, tasteful, confident. Basically, more cool. When did my infatuation begin? When I bought Rubber Soul and fell in love with John Lennon, of course. Wasn't every girl my age in love with John Lennon? Which means, I guess, that mammals are to dinosaurs what I was to John Lennon. A scrap of nothingness.

~~~

If we zoom skyward a moment (hold tight), we can see dinosaurs on every continent. How did they cross the oceans? They didn't need to. When they first appeared two hundred million years ago (194 million years longer than Homo sapiens have been around) there was just one chunk of land. But look at them now. Five whole land masses, each evolving their unique ecosystem with their dinosaurs evolving in different ways too. It's not over, it's still happening—two of these dinosaur-laden continents are breaking up right now. If we speed up time briefly, you can watch them gently slide over the surface of earth. They're sliding as I write. (I said, *hold tight!*)

But none of this matters to Ollie the olorotitan. I'm not going to tell him his bones will leave an imprint in mud which will become stone, or that Russian scientists will pore over it and spend decades discussing him. I mean, look at him—obviously shy, doesn't want attention. I'm whispering these words David Attenborough-ishly. Don't want to alarm Ollie.

Look at him.

Just a toddler, really—considering he'd live another eighty years if disaster wasn't on the cards. He's got a weird duck-bill-shaped thing growing on the top of his head, straight up from his snout—no doubt quite attractive to other olorotitans. To his mother, anyway. He's grazing on some trees, methodically stripping the leaves from each branch and calmly munching away. It's a rhythmic process. I feel mesmerised. Tug, tug, rip. Chew, chew, chew. Tug, tug, rip. Chew, chew, chew. His nose is huge and crest-shaped and he can make an alarming trumpet noise with it. How do I know? It's how I found him—I followed the noise for an hour and there he was, trumpeting for the sheer hell of it. He's not the only one. I can hardly hear myself think, there's such a racket going on. His long tail is down, relaxed, and he doesn't seem aware when it snags on some jagged bits of outcrop. He has some cuts and scars, presumably from the accidents common to all toddlers. From end to end, he's about the length of three cars. In short, Ollie is a fine specimen of baby dinosaur-hood.

Okay. I think he's spotted me now. He's gone still.

'Hey, Ollie,' I say quietly but looking straight up into his large green eyes twenty feet above me.

He jumps as much as something his size can jump. Because I'm so near, it feels like an earthquake.

'How do you know my name?' He has a high pitched girly voice. It's kind of funny, coming from something so scary looking.

'I don't. You don't have a name. Ollie is a nickname I gave you from your species name. The truth is, I only have access to the names of direct ancestors.'

'I don't get a name?'

'Nope! Guess you're not important enough.' I say this in a teasing way, and hope he understands the opposite is true.

'Am so.' He stamps one foot and lifts his head.

'Who says?' I reply automatically, although I'm confused. Is *he* teasing *me* now? So hard to read facial expressions from this distance. All I can see now is the underside of his chin and the insides of his nostrils. Some crusty stuff has dried on the rims.

'My mum, that's who.'

'Well, what does she call you?' I say, in what I hope is a kind tone, then I step slowly backwards so I can look him in the eyes again.

Pause.

'None of your business.'

I don't remind him that he asked how I knew his name, and that logically therefore Ollie *is* his name. Maybe he's in a bad mood and will contradict whatever I say, which reminds me of one of my daughters when she was ten to seventeen.

I'm starting to think he's not shy at all. He's confident enough to argue, which means he's cared for. Looked after, fed, kept safe, told he's special enough not to kowtow to any Tom, Dick or Harry. I imagine his mother as a firm no-nonsense kind of parent. Ambitious for her son. My mother wasn't like that. She was wishy-washy and soft and funny and thought I was a genius if I tied my own laces at age twelve—but then she didn't need to prepare me for survival in the Mesozoic Era. I suspect successful species at this stage of the game depend on the firm no-nonsense kind of mothering. Actually, my father was more fit for the Mesozoic era. He liked to toughen us up by teasing and sometimes becoming the tickle monster. Those were scarring episodes.

'Did you hear me?' says Ollie. 'I said *none of your bee's wax*.'

'I'm sorry. Truly.'

'What do you want, anyway?'

'Not much. I just want to watch you, Ollie. I'm curious. You may not be one of my direct ancestors, but we're still related. Sideways, like cousins. We have some of the same ancestors from 435 million years ago.'

'Yeah?' Still a bit huffy.

'Yeah. The main guy was called Kevin.'

'So why does he get a normal name? Or was his species called Kevinsoraus?'

'You're missing the point. Aren't you interested in Kevin? I think you would have liked him. Bit inexperienced and simple, but a very strong personality. Open to sharing.'

Ollie kind of *harumphs*, swishes his tail and puffs out his chest as much as a dinosaur who hates sharing can. As he does this, he creates a micro climate. Everything in his vicinity is poised, waiting to see if he's finished harumphing now.

He's so different from the others, I think to myself. Kevin,

Polly, William, Betty, Margaret and Beryl were all so much, well, less. I can see why dinosaurs, like Neanderthals later on, are the celebrities of their time, though neither species is required for us to evolve. Their extinction won't matter to you and me. Imagine an uncle dying young. Tragic, but it doesn't mean your parents won't still end up making you. You're safe. And yet dinosaurs are easily the most famous prehistoric animal. We can't seem to get enough of them, especially scary T-rex or friendly shrub-munching triceratops. We're less sure what our ancestor—Harriet the Forgettable—looks like at this stage. Various teeth and fossils indicate she's rat-like, close to a shrew in appearance. Probably nocturnal, and she probably has a tail—as did you and I when we were in the womb. A souvenir from our prehistoric days, which disintegrated before we were born. Harriet belongs to the species which directly leads to animals with placentas—of which we are one. Thanks to Harriet the Forgettable, our babies can remain inside us longer, giving us plenty of time to buy baby-grows and books with titles like *How to Not Do it Wrong*.

How did we get from six foot therapsid Beryl to dinky Harriet? The same way we got from one cell Kevin to multi-celled perky Polly, only backwards. Slow change over immense periods of time can result in life form expansion or shrinkage—depending on which size works best in terms of survival. It seems our ancestors who functioned best between Beryl's time and Harriet's time were increasingly small, until voila! Harriet. Who is quite small, even for a shrew—in fact she's probably a pygmy shrew. Larger animals will need too much fuel to survive the next catastrophic event. She's easily the most important creature here in terms of Homo sapiens later evolving, but—bless her heart—how much charisma does she have compared to Ollie? Who'd pay £15 to watch a film starring an inconspicuous shrew living modestly? Who'd buy an over-priced soft toy afterwards, inspired by a main character called Harriet the Forgettable? The answer is *nobody*. Harriet is humble, and humble might endure—my self-effacing mother lived far longer than her bad-boy husband—but it does not sell. Even me, her grateful direct descendant, would rather spend time in the company of Ollie.

'Sorry, Harriet, for ignoring you. If you're here.' I whisper softly so Ollie can't hear—I'm guessing he's a jealous guy.

I tune into the noises around me, of which there is a constant abundance, and hear the tiniest of squeaks. Squeaks of forgiveness? I might have imagined it, but I whisper back:

'Thanks, Harriet. Very generous of you. And by the way, happy birthday!'

Another series of squeaks, which I interpret as: 'Where's my present?'

I place a tiny cupcake on the ground in the vicinity of the squeaks. It's made of mashed up leaves and ants with a compostable candle on top. She drags it into the undergrowth so quickly, all I can see is a blur of brown. No lurid colours for my classy Harriet! I hear a rustling noise followed by a swallowing, which I assume means she's eaten the whole thing.

'Was that tasty?'

No answer. Not into small talk, I think. That's fine. I focus on Ollie again.

'Ollie, listen—I was just teasing you before. You're so famous it made me nervous, and my father taught me: *When in doubt, tease defensively*. But the truth is, I can hardly believe I'm talking to you,' I gush like the true groupie I am. 'God dammit, Ollie. You're the coolest thing on the planet.'

'You said a swear.'

'Sorry. I get over-excited when I'm excited.'

'Okay,' he says, swishing his tail. 'Actually, I knew that already. I'm the coolest thing on the planet and I know everything.' Which is a lie and just means he hates not knowing stuff. He sniffs a few times, as if he's trying not to cry.

'Of course you knew. I brought you something, Ollie.'

Then I give him a small tree I've lugged from the botanical gardens back home.

'There you go. Enjoy.'

I feel kind of sad, watching him devour the greenery. He's so happy now, oblivious to everything but eating a gourmet tree. All he wants in life is to eat like this, maybe meet a nice girl and make a bunch of little olorotitans one day. He finishes the tree and starts plodding along. I follow at a safe distance, and watch him trip several times trying to avoid smaller creatures scurrying in the undergrowth—one of whom is my Harriet.

Keep dodging danger, Harriet! I think, and immediately hear a two-syllable squeak in response. I have the grace to acknowledge her modesty, and think: *You're pretty cool too. I really respect your self-sufficiency.* She's not sulking because I gave Ollie a whole tree and her a minuscule cupcake. I give her what I consider a respectful salute—slow, serious—but if she's still there in the undergrowth, she's not wasting any more squeaks.

~~~

Ollie may be arrogant by comparison, a little insecure and prickly—but he's done nothing to deserve what's about to happen to him. Nor have any of these grand and glorious reptiles. Nor the technicolour fish, or the three eyed mammals, or the fluorescent insects, or the bizarre and lush plants—which I keep forgetting, despite the fact they too are life forms with cycles and complex cellular structures and plant versions of circulatory and respiratory systems. Some of the ones Ollie eats can defend themselves from parasites, and others are carnivores. That one just above you can communicate danger to other plants, just like Ollie's mother trumpeting to him when she senses a stegosaurus stampede.

Look at it all.

Buzzing, shining, thumping, squelching, blossoming Life. The whole shebang is just going about its day, trying to stay alive and be a good citizen. I look up, hoping it's not there—maybe I made a mistake—but there it is. Can you see it? Small, but unmistakable. A celestial body—an asteroid or a meteor or a comet—is hurtling towards Earth at the speed of 40,000 miles an hour. Sometimes foresight is a curse. I'm glad Ollie is ignorant.

The world has probably never looked so beautiful. The light in the trees and the colour of the flowers and the shadows moving over the wetlands. The smell of dinosaur poo and sulphur from the springs and something sweet, very like frangipani. *Listen*. A hundred thousand birds and animals and insects and plants are saying they're hungry, delighted, cranky, broody, afraid, even just bored because this is a day like any other. Most of them are about to die, which is sad. Then again, *thank God*. Because otherwise they'd continue being the dominant species, and this world would evolve into a reptilian version of our current society. Harriet's descendants (us) would be hunted to near extinction, then exterminated just to tidy up the place.

'Ollie,' I say softly, because to speak at all is hard now. I stretch the syllables out lovingly. Sing his name. '*O..lee!*'

'Hum?' His mouth is full again.

'Can I touch you?'

'No,' he says. 'Not allowed. Mum said.'

'Okay,' I answer.

My heart is heavy, but I smile as I turn to leave. Give him a little wave and call out a jolly goodbye as if I expect him to live happily forever after. He doesn't look up.

Under my breath, I mutter: 'Hold tight, Harriet.'

Two squeaks. I think she means *I will*. Or maybe she's asking: *What for?*

~~~

It would be very easy to feel smug right now, making my way easily back home where a celestial body is not about to crash. Maybe it's hard not to feel arrogant when you're a member of the dominant species on Earth. The alpha race—that used to be dinosaurs, and now it's us. Then I remind myself I'm part shrew. It's hard to be a snob when you accept you're biologically part-shrew. I turn around one last time and blow a kiss to Harriet.

Jeff

14 million years ago
32 miles north of where Barcelona will be

So much has happened post celestial body impacting on Earth, and not all the outcomes have been bad. *Dinosaurs are dead, long live the descendants of Harriet the Forgettable!*

I know. Doesn't have the same ring does it? Maybe evolution is not a logical linear process after all—maybe it's more a series of accidents, both lucky and not. 'One step forward, two steps back,' my mother used to say whenever there was a hiccup, completely unaware Lenin had made up the phrase. She had an allergy to communism and would've hated knowing that. It's been a while since Harriet scurried into the undergrowth—51 million years ago, roughly. The comet/asteroid/comet disaster led to the biggest mass extinction event since…the one before that. There's been three, and counting, as well as the slow constant fluctuations in climate that have beleaguered Earth since it was a hungover woman trying to remember her name. The celestial body killed 70% of land animals and 90% of life in the oceans, but not our hardy Harriet! No, no no!

Meet her grandson times a lot—Jeff. Not much resemblance, is there? I mean she was a tiny shrew-like animal and Jeff looks more like a monkey. Both have body hair but that's as far as the resemblance goes. Harriet was an early mammal who evolved into a Jeff.

'Hey Jeff. Jeff. Jeff!'
'Jeez. What?'
'Say hello to our visitors. They've come all the way from the future.'

'Do I have to?'
'Yes.'
'Hello.'
'I can't hear you. Speak up.'
'HELLO.'
'Thank you. How are you?'
'How does it look like I'm doing?'

Hmm, I think. A bit Eeyore, like William the worm, plus cranky like Beryl the therapsid. I can handle this. I've raised four rowdy kids, I'm built of steel. Don't react aggressively, I tell myself. Save time and pander.

'Okay. Not so good?'
'Correct. I'm hungry. And fricking freezing.'
'Ah. Yes. The planet's been steadily cooling for a million years, and you've not evolved sufficiently to guard against lower temperatures. Your descendants will fare better.'
'What? Who?'
'Us lot. The life forms who come from you.'
'Huh. Do you have a fur coat I can borrow? I mean have.'
'Sorry, no.'
'Hand warmers? Hot water bottle? Thermal underwear?'
'Nope. Damn, that would've made a great birthday present.'
'For who?'
'For you. It's your first birthday today. Or the equivalent of.'
'Can I have that thing you're wearing?'
'My cardigan?'
'Correct.'
'No.'

~~~

Poor Jeff. Climate change is not his fault. Maybe I'll tell him it's going to get worse, much worse, and he should count his lucky stars because he'll not witness the worst of it. The ice sheet in

Antarctica is growing rapidly, but the next proper ice age has not begun officially. Although, in another way, it's beginning right now in Jeff's time. Hence the shivering and begging for a fur coat. (It's easy to date something ending, not so easy to date something beginning. There's always something before that which leads to the thing happening.)

'What's that sound?'

'Oh, nothing,' I say nonchalantly. 'Just the sound of trumpets faintly heralding the advent of Ice Age Number Five.'

'That's not funny.'

'I know.'

'Can they play happy birthday?'

'I don't see why not.'

(Sound of trumpets playing happy birthday.)

'Aw, thanks. That was nice.'

Quite sweet, the way his eyes have softened. Maybe no one has made a fuss of his birthday before. We're in what will become Spain and it's winter. Jeff is a Dryopithecus, unless he's a Ramapithecus—no one is quite sure which is our direct ancestor and which evolves into apes. Not all primates lead to Homo sapiens. We do know this about him: When he chews, his jaw grinds side from side. He's got very little hair on his face and hands, and his body hair is reddish and quite long. He feels most at home in trees. When he grows up, he'll be about four feet tall, or four feet long, and weigh twenty-five pounds. His front arms are disproportionately long. Maybe it helps with swinging through treetops. I was watching him play on the ground earlier, and I noticed he ran around on his knuckles until his mom called *Dinner time!* Then he briefly stood up on his hind legs before reverting to the knuckle walk.

'What's for dinner, ma?'

'Guess.'

'Fruit again?'

'Yup.'

Which explains his weak cheek teeth and weak jaw. He doesn't need to chew tough roots or meat. His face is actually kind of cute. Like a chimpanzee, but only a little. (Chimpanzees will not appear for eight million years.)

'Come here, honey,' I say on impulse.

I pick him up and let his long arms curl round my back like a baby, only a longer reach. His legs straddle my waist, and he strokes my face with fingers like black leather. There is no tenderness in his eyes—that would be sentimental—but there is curiosity. Because he's spread himself evenly over my torso, it's easy to walk around with him, which I do. I let go with my arms and he stays put. I used to do this with my kids who were all skinny-malinks and easy to carry for years past babyhood. I loved that phase, though I complained a lot at the time.

I unpeel Jeff's limbs and put him down.

'Reach for the sky,' I say and he does, clever kid.

I slip off my cardigan—size 14, mustard coloured from White Stuff—and try to put it on him. It's hard because it's way too big, and cardigans are designed for upright two-armed life forms, not little fruitarian tree dwellers. But he's cold and I'm fond of him now. He's made me nostalgic for early motherhood, so I find a way. In the end he looks like a mummy, but at least he's stopped shivering.

'Do you have a girlfriend?'

'I'm one year old.'

'Sorry.'

I look around, see if I can spot my grandmother times a lot. There's a Dryopithecus on the lowest branch of that tree by the river.

'She's pretty, right?'

'Too hairy.'

'Oh Jeff, I think hairy might be a good thing, now the weather's getting colder.'

'Huh. Never thought of it that way.'

'Say yes to hairy girls, Jeff. Honestly. Think of the long run.'

Then his mom hollers out that dinner is ready. Again, I think? How long have I been here? Are my cookies burning back home? My grandkids are due off the bus soon, and here I am dilly dallying with the past again. I try to hide a bout of self-loathing, but no need. Jeff has already scrambled off, into his global-freezing future.

'Hang in there,' I say to no one and head home, hugging myself. I wonder how long my cardigan will last. Fast fashion items are not renown for longevity. As for surviving an ice age, not a chance.

Freddie

**Six million years ago
What will be Africa**

Things are finally starting to speed up on the evolution front. An acceleration has occurred, or is occurring. An accelerating acceleration. Eight million whole years have gone by since darling Jeff the Dryopithecus's time, and here we are! If Life is a party, hominids arrive so late only the inedible nibbles remain. Dry triangles of pizza that someone has picked all the pepperoni off. Empty Pringle tubes and crushed up beer cans. *Pour yourself some flat prosecco and mind that puddle of puke in the hall.* Ignore me—I'm just being negative and ungrateful.

Take a deep breath.

Feels different, right? Oxygen has increased since Jeff's day, and doubled since Harriet's. This is due to the increasing vegetation—mainly blue-green algae which is basically a population-increasing machine, replicating every nine minutes. It's especially good news for the hominids, because bigger brains require more oxygen. Low oxygen equals low IQ, although that implies stupidity lessens chances of success as a species—and this is clearly not true. Behold William the worm, who is thriving still. Or ants, who've already been around for more than a hundred million years, through ice ages and asteroids. Or more worryingly, behold viruses—like us, stemming from Kevin, although one or two scientists think they might be the precursors of Kevin and the driving force behind all evolution. Imagine! Alien viruses who insist we host their life support systems and force us to adapt or die. Yes, big brains serve us well, but our biggest threat may be brainless micro-organisms who don't know the meaning of the word extinction. They are the stuff of nightmares. Let's not talk about them.

Let's talk about hominids. Freddie is a hominid. (Albeit ridiculously named, even by my standards. Life knows no bounds when it comes to cultural inappropriateness.) Here we are in Hominid-ville, East Africa. The Rift Valley in Kenya, to be precise. There are other groups of hominids scattered around the world right now. Among them, the Neanderthals up in Europe and over in western Asia. The Denisovans, also in Asia. And perhaps most intriguingly, the three-feet-tall Hobbits in the Indonesian island of Flores who co-exist with pygmy elephants and Hobbit-eating lizards. They will exist a surprisingly long time before biting the dust like all the other hominids except for clever amazing us. Or rampaging colonising us.

Look around.

Yes, yes, I know. It doesn't look much like the Kenya you saw in 'Out of Africa', does it?

For one thing, it's not green—it looks like a place that used to be green and which is now on the verge of desiccation. Just the sight of that dry riverbed over there makes me thirsty. We're still in the interglacial period that Jeff lived in, leading up to the fifth ice age—the one we are technically still coming out of in the twenty-first century. Our ice age properly began about three million years ago when an emerging isthmus stopped the Pacific and Atlantic oceans from talking to each other, hence cutting off a warm current. We are all—Freddie and us—in interglacial periods with a big fat ice age in between us. Greenland is not green, though that was never why it was called Greenland in the first place. Blame naughty Eric the Red trying to entice settlers—one of the earliest cases of fake news. Antarctica is likewise entirely covered in ice during Freddie's time, and all land masses near the poles are suffering low temperatures. But here in northern Africa, the climate change means hotter temperatures and far less rain.

Why are we here?

To celebrate Freddie's first birthday, of course. He's Jeff's grandson roughly times a trillion. He's what Darwin was talking about when he said: *Man is descended from a hairy tailed quadruped, probably arboreal in his habits.* Freddie's species of hominid will last about 5,800,000 years and end up in you and me. Homo sapiens.

'Hello there, Freddie!'

I have to bend my neck right back to see him. Can you see him too? He's straddling one of the lower branches of that tree which has furry yellow balls hanging from twigs like Christmas ornaments. He's entirely covered in fur, but don't be fooled. We have exactly the same number of hair follicles on our skin as he does. Our hairs are just lighter and shorter.

There's others like him nearby, but they're bigger. He's busy eating something.

'Yoo hoo, up there,' I call out.

He's ignoring me, but that's alright. I'm never sure how ethical it is to interact with members of extinct species. I always tell myself to observe, take notes, contemplate, be objective—but then I open my mouth and out pop birthday greetings and nosey questions. Or in this case, out pops a lullaby.

'Rock a bye baby, in a treetop, when the wind blows...'

It's gratifying just being in his presence, which means I'm kind of a racist. I prefer beings who resemble me. And compared to a worm or a penis-shaped microbe or a shrew or a fish or even darling Jeff, Freddie is hugely relatable. Actually, if you like kids, all chimpanzee-like things are relatable.

'Freddie, honey...' I croon.

Still no response. That's okay.

He really is one year old, not a primate equivalent. I imagine, if he notices me, he's quite unimpressed. And not startled, because all things in a toddler's world are equally unfathomable and therefore mundane. (I'm starting to have days like that, so presumably we get to experience that twice.) I begin to sing Clementine, and finally Freddie stops eating. He looks straight down at me with calm dark eyes. He spits out a seed that lands on my nose and I leave it there. I don't even twitch because that might seem rude. Maybe the seed is a gift. I walk closer and his eyes follow me. I can tell because there's white around his irises. Not sure I've seen the white of an eyeball so far in evolution. It's uncannily nice. I feel properly perceived.

'Freddie! Freddie, can you see me alright?'

'Of course I can. I've got two eyes, don't I?'

Why are a lot of my ancestors cranky? (Why am I so cranky? Yes, yes, I rarely act cranky, but that doesn't mean I don't have cranky thoughts, dammit!) Also, why does he have an English accent and sound like Stewie in Family Guy?

'Sorry.'

'What do you want?'

'Oh, nothing really.' I feel shy suddenly. A little scared, even. It's definitely easier talking to pre-hominids. Creatures I feel superior to.

'I didn't mean to bother you.'

'Too late. I'm bothered,' he says, and grabs another fruit.

I note his grown-up tone, and think: Of course he's not a typical toddler! He's going to change the whole game.

Look at him.

One thing's for sure—he may look a bit like one, but Freddie is not a chimpanzee. That's because they still haven't evolved. Nor have gorillas or bonobos. But we share something important with chimpanzees, which explains why our DNA is 99.5% the same—we share Freddie! Who is not a chimpanzee or a Homo sapiens, but the thing we both come from. An early hominid. If you compare him to the others in the tree—and I mean family tree, not just the yellow-ornamented tree he's sitting in—his neck is longer and he can swivel his head more easily. His forehead is noticeably higher. Look at the others, in both trees. Most don't even have foreheads, their faces end at their eyebrows. He grins down at me, and I use my handy binoculars to note that his teeth are smaller than his grandfather's (who I recently met), but the enamel on them looks much tougher. In fact, they're very like mine, and I don't think that has happened in evolution before either. I study his hands. The tiny digit on the right of the left hand and on the left of the right hand is interesting. It's been evolving for a while now, but this is the first time it's functional enough to warrant the word thumb, not just fifth finger. It helps him grasp delicate things and more easily peel fruit, crack nuts and pick up insects. Think that's a no-brainer? Try to do anything without your thumb cooperating. Go on, try it. Forget using a pen, or a fork, or pulling on a sock.

'What do you want?'

'I'm here to celebrate your birthday, Freddie. It's worth celebrating because you're my grandfather times a million or so.'

'That's a lie. Where's your fur? And what happened to your face? No offence, but I can hardly look at you. Ugly.' He says it like this: Uh. Ga. Lee.

'Believe me, it's a biological fact that a tiny bit of you is inside me.'

'Right now?'

'Right now, as we speak.'

He stares at me, his mouth slightly open. I'd like to say he has an intelligent expression, but right now he just looks just a dumb kid. In fact, some drool is spooling down his chin. I wait for him to respond, to ask something else—but he shifts his gaze to somewhere past my left shoulder, and his eyes get an unfocused content look, like he's suddenly daydreaming about lunch or contemplating his next bowel movement. He makes a grunting noise three times.

'Oom, oom, oom!'

He sounds just like a monkey. Also a boyfriend I had briefly a long time ago who was from Tulsa and drove an eighteen-wheeler.

'Oom, oom, oom!' I grunt back, to show solidarity.

Freddie rolls his eyes. Maybe I'm trying too hard.

~~~

At first I thought his name might be George. I was thinking about a fictional monkey who was popular when I was a kid—Curious George. Like George, I'm guessing Freddie is also curious, and that curiosity will save his ass. I put my binoculars away and pretend to be interested in the fauna around me. Sniff it, feel it, take a few photographs. This is the first time I've brought my camera. It's probably against the rules, but who's going to report me? I'm about to click the shutter when Freddie looks at me sharply.

'No pictures.'

'Oh. Right. I forgot.'

'Sure you did. Put it away.'

'Who's your best friend, Freddie?' I ask casually, to change the subject. I used to ask my kids the same question, it's natural kid small talk. Almost every day there'd be a new best friend, but I'd ask for the name anyway and what they looked like and if they both preferred Jelly Babies to Kinder Eggs.

'I don't have one.'

'Oh, never mind. Best friends aren't a big deal. Are your friends in this tree?'

'I have no friends.'

'You have no friends?'

'Jeez. Rub it in, why don't you.'

So, not a popular kid. Poor Freddie. Maybe it's his higher

forehead, which looks odd enough to earn a degree of social shunning. They're not carnivores, his family, unless you count eating ants and fleas. They've survived so far by blending in, keeping their heads down when there's danger, and by not thinking very hard. In evolutionary terms thinking is over-rated, even counter-productive, because the brain requires energy and food is never consistently abundant. Besides, thinking can result in discontent which is only helpful if it gets you out of a jam. Not good news if you are a worm or a shrew, which is why they're blissfully small-brained. But here we are with Freddie—a big brained fledgling hominid in a family of smaller brained fledgling hominids. In Hominid-ville, it's probably the law to keep the species pure. To not breed if you're different. Oh God, what if their credo is to *kill the weirdoes with big foreheads*? Poor no-friends Freddie.

'It's okay to be different, Freddie. I like the way you look. You're handsome just the way you are.'

Now I sound like Mister Rogers from the American children's tv show, but I'm sincere. I do like the way he looks. The rest of his family, to me, look more like pseudo chimps.

~~~

Why has Freddie been born slightly different? I could make up a story about a minute chemical shift during his formation. Or the berries his mother ate the afternoon when she was in early pregnancy, which resulted in hallucinations and then vomiting. Or the iron-rich meat she loved to eat on the sly, little rodents crammed into her mouth when she thought no one was looking. Maybe it was the lightening which struck her briefly later that month and accelerated foetal brain development, though that sounds too much like Frankenstein. Freddie is not a monster. He's a little boy monkey-thing, sitting on branch, watching the other little boy monkey-things play games without him.

Any evolutionist would say all my ideas are piffle because environmental factors cannot affect genetic material. The truth is much more basic. All genes mutate (change). Change is the only process that never changes. Sometimes it's good news for a species, sometimes not so good. After a long period of time, one way or another, the mutations cause a new species to form. (What defines a new species? A generation becomes so different it can no longer breed with its own species, only other newbies.) Freddie's bigger brain, an evolutionist would probably say, is simply the result of millions of years of experiment and accident—perhaps driven by a subliminal need to keep the species alive. The smarter ancestors with the bigger brains kept staying alive longer and making more babies, and the smaller brained ones started dying out. It was so slow a process no one noticed the change at all. It's as undramatic as that, this respected evolutionist would say, maybe clapping his or her hands together for emphasis. But if this is true, why is Freddie the only big-brain around here?

Does anyone know?

Richard Dawkins, one of the more popular theorists, said it's all about the genes themselves, not us. Our genes existed before us—some of my DNA is 3.5 billion years old—and they are hell bent on living forever. Our whole life we are manipulated for the gene's benefit, and we're just kidding ourselves imagining any different. It's not social responsibility or God's grand plan that keeps us going, it's each gene striving for its own sweet self. If we fail to reproduce, it's basically the gene committing suicide or being left to die, bitterly gasping something like *You let me down, you bastards!* I'm inclined to think Dawkins might have a point, mainly because I like his unpretentious writing style. But if he's right, what am I do to with Freddie, who is indeed a little odd looking.

'You're bananas.'

'What? Sorry, did you say something?' Maybe he's been talking for a while.

'I said you're bananas,' says Freddie. 'It's none of those things. I'm different because I'm different. No reason.'

'Oh, come on,' I say, a little peeved. This is my story, after all. It's annoying when ancestors have opinions.

'It's your problem, not mine. This allergy to mystery. Why can't you just leave it?'

'I'm curious.' And I'm not George, I say to myself.

'You're nosey. Does knowing stuff make any difference to anything?'

'Well, yes!'

'Like what? If you knew why I'm evolving into another species, why my brain decided to be bigger, how would that change anything?'

'Oh! Well, that's easy. It would change how we...I mean it would alter the way I...'

Big pause. There's a sound in the distance of something like laughter, but that's probably me being paranoid.

'See? Pointless.' Then he smiles kindly, as if he's humiliated me—which he kind of has—and tosses me a piece of unidentifiable fruit. I conceal it in my closed hand and pretend to eat. I don't want to offend him, but I don't suppose his hands have been washed properly in his entire life.

'Nice,' I say, as I pretend to swallow. 'Yum.'

'Just toss it on the ground if your hand's getting sticky.'

I let the fruit drop and a striped green reptile scoots in and steals it.

'Listen, nothing stays the same. Things change!' says one-year-old Freddie in his camp Stewie voice. 'You think I like being called Fat Forehead? Being a freak here? But we don't have control over the way we change. It happens.'

'Yeah. Thought as much.'

'Sure you did.'

~~~

This day is not going the way I thought it would—a perfect case in point. Change is the only thing we can be sure of, and even then Life bursts the banks of our tidy expectations. Maybe Life itself, seeking endlessly to live, causes every living creature to keep evolving. Logically, Homo sapiens should be improving, although lately I've begun to wonder if the signs of our extinction are on the horizon. Has anyone measured the typical human head recently? Apparently, our brains have begun shrinking. Maybe we're evolving into two sub-species of Homo sapiens, just like one particular group of hominids evolved into Homo sapiens and chimpanzees. Maybe in a million years, we'll be known as the Big Foreheads and the Little Foreheads. That would explain a lot, but it's also worrying because bigger brains require more fuel, and if Earth can't feed us all, well! Not good news for the Big Foreheads. Naturally I put myself in the Big Foreheads category, even though I know this means possible starvation and scorn from the Little Foreheads. *Arrogant cow! Thinks she's better than us!* That's how much I am conditioned to think intelligence is the best thing an animal can possess. (*Yes, yes, William. I know it's untrue!*) I sit down under the tree because I'm very tired suddenly.

'Thinking too hard again, I suppose,' says Freddie with a chuckle. 'The harder you think, the more energy you use.'

'Jesus. How do you know so much?'

'I'm the smartest cookie here, that's why.'

Then he swings to the ground and sits next to me. He smells of fruit sugar, Labrador puppy, and weirdly, a splash of gamey popcorn. At first I don't like it—but within seconds, I do.

'Listen. Something's happening to the climate here,' he tells me in a confidential way, like he's repeating some juicy gossip.

'It's not a good pattern, and I feel like I should be doing something about it.'

He explains it's been getting hotter and more worryingly—drier. There's not enough rain, not even enough humidity, which means increasingly fewer fruits to eat.

'But you're only one,' I say, mostly to myself.

'You think I'm stupid? I know what's going on. The world is coming to an end.'

The rest of the world will get hotter soon too, but for now, equatorial places are worst places on Earth to live. Very soon it will be untenable in survival terms. I don't tell him this because I mustn't influence him, though I *will* him to consider a nomadic lifestyle soon.

'Gosh,' I say. 'That does sound serious.'

'Not sure the others have thought about it much.' He indicates his family with his eyes, which only works if you have whites in your eyeballs. I begin to think of all the ways in which eye whites are useful, when one of his aunts swings down, squats behind Freddie and starts picking out his fleas. I'm so relieved with this proof that he is not really an untouchable, I almost hug her. I note she doesn't smell as nice as he does—a complete absence of Labrador puppy and her teeth are a disgusting furry brown. Freddie holds perfectly still. Then he finds a flea on his aunt's thigh and pops it in his mouth. For a vegetarian, he seems to quite like blood. There they sit, aunt and nephew, gorging on each other's fleas. Fleas are already old hat. Dinosaurs came and went a hundred million years ago, but fleas go on and on. Freddie still glances at me from time to time, but his aunt is oblivious. Then suddenly, he stands up.

Freddie. Stands. Up.

It's a big deal. This is the first time in over two billion years I've seen an ancestor properly stand up. (Jeff could do it for a few seconds, but that was all.) I stand up too and he comes up to my thigh, maybe thirty inches high. My guess is he'll probably grow to four feet tall and live about thirty years. He takes a few steps, arms swinging, hands nearly touching the ground—all skinny arms and legs and potbellied and big headed—so cute, I could gobble him up. I follow him to a nearby bush sporting red oblong things. Some insects are having a hey-day there, but Freddie disperses them by blowing hard. He settles on his haunches and commences peeling the red fruits. They're mostly dead looking and wrinkled, but he doesn't appear to care.

He turns and glares at me.

'Do you mind? I like eating alone.'

'I do too sometimes. A decent brownie, for instance, is always better in private.'

'And?'

'Okay, I'm going.'

I walk back to his aunt, who yawns when she sees me coming. It's a very long yawn, and I get to see her many molars. Wisdom teeth are remnants from our primate past, when we needed extra molars to chew tough things like roots. I'm just about to ask her if she ever considers brushing them with twigs and leaves, when she starts chatting to me about Freddie. (The only other non-ancestor who has been able to see me was Ollie but I pretend I'm not shocked. Maybe she's a great grandma. Or maybe Life's software has a virus.) I learn Freddie is not an early walker—in fact his siblings and cousins all walked a month before he did. Nevertheless, I want to tell her, Freddie is *the one*. (I don't, of course.) Then she tells me that Freddie's mother had another baby last week, and both baby and mother died. Apparently Freddie lay by their bodies for several hours.

'Oh dear! He must have been so sad.'

'Sad? It was a chilly night and their bodies were still warm,' she says, shrugging.

'Oh,' I say, making a mental note that grief might be a luxury reserved for the affluent future. *No love yet*, I think. Or no Love with a capital letter.

'It wasn't that he didn't love her,' says the aunt.

Jesus, you can read my mind too? I say silently, and she simply nods in reply.

'She loved him too. They were always kissing each other, those two. Us hominids are excellent mothers. If she was alive, she'd nurse him for another four years.'

She sounds so proud I half expect her to take out some cute baby photos.

'So, he *was* sad?'

'Of course, but that's not why he stayed close to her body. Aren't you listening? Death is common. Common as…as sneezing. It would be uneconomical to waste time and energy grieving when we could be food gathering. Personally, I've got to keep my milk coming or at least three of the babies won't make it to winter.'

'Ah. She was your sister?'

'Maybe. There's no way to know, but maybe. Aunt is a courtesy title. I nurse Freddie these days, so do a few of the others. If I hear him make his hungry tired noise, I just scoop him up and plug him on. Anyone looking would assume I was his mother.' She touches one of her nipples contemplatively as she says this.

'What about his father?'

'What's a father? You mean the big man? That's always changing. There's always a scary fight first. I try to herd the little ones away when that happens.' She sighs, but I can't tell if she's signalling impatience with the alpha male ritual, or with me and my questions.

'You sound very…very wise,' I say eventually.

I look over at Freddie still eating his red oblongs, his face and chest covered in fruit juice. He's completely absorbed in sucking the flesh, spitting the seeds, then chucking the peel. A small pile has grown to his right, and I wonder if this makes him right-handed, or if such a thing exists yet. Now I know he's essentially an orphan, I feel even more motherly towards him— even though I know if I stole him home with me, within three nights I'd be pulling my hair out. Broodiness is a con job. Maybe Love is actually Life conning us so we keep the whole show going—because the result is never anything like the promise. Not to say I wish I'd never had kids, just saying motherhood was a painful surprise from time to time. Nevertheless, what replaced the dream was better—Love with a capital letter. Imperfect, ragged, adrenalated, mundane, contrary Love. I *love* Love! I give it to my kids and husband, and then actively look for other recipients to pour it into. The well is never dry, which is lucky because people irritate me intensely sometimes. Okay, often.

Now I know that Love with a capital letter is apparently here, I wonder when it began and why it began? What is the evolutionary function of Love? The physiology of love is easy to observe and quantify—both romantic love or lust (high blood pressure, genital arousal via engorgement, decreased rationality) and platonic love (lower blood pressure, healthier immune system, increased sense of security). But is Love essential to survival in the way, for instance, Fear warns us of danger? Maybe beloved people live longer. And maybe sometimes lovers are like hunters who aren't very hungry. Who only really enjoy the act of hunting, and then leave the dead animal on the forest floor.

'You're right about that,' says the aunt, rocking a little on her heels. 'I am pretty wise.'

'And we're probably related, so that makes me wise too, right?'

'No.'

Then she tells me another story about the time clever Freddie figured out how to do something or other. Having a high forehead might mean you have no friends, but you're still loved by the mothers—which is lucky for Freddie. I feel such respect for his aunt. She mothers him out of the goodness of her heart, and lord knows mothering anyone takes a lot out of a person. Then again, maybe maternal Love isn't as voluntary as we think. Maybe it's compulsive. Maybe the ancestors who had more selfish independent tendencies tended to die out due to their increased vulnerability. No pack to hide in, no family to protect them. Like it or not, and some like it less than others, being a wimpy social creature is safer. Though the leaders—the more selfish individuals who don't crave safety or approval—often shape our history.

'Me and you,' I whisper to Freddie's aunt, 'we're not destined for much. Our lives will not alter the course of humanity.'

'Bummer,' she whispers back, and looks genuinely disheartened.

'But we're good mothers,' I hasten to tell her. 'We're not nothing.'

'You're just saying that to make me feel better.'

Where have I heard that before?

~~~

Fast forward a year. (Travelling into an ancestor's future beyond their first birthday is a benefit limited to hominid and Homo sapiens ancestors. More proof of ethnocentricity. I'd feel guilty, but that would be hypocritical because I'm enjoying the entitlement too much.) The drought is getting worse, and Freddie's intelligence—housed in his freakishly large head—tells him a journey is his only hope. It says *Go west young man*. Or rather, north-west. He doesn't plan his journey, not consciously, and certainly doesn't tell his family or take anything with him. Oh no. He leaves home rather like an adolescent reacting to hormones and also a nomadic gene from his mother's side. They were always a shiftless lot. Like a sixteen-year-old slamming the door and shouting: *I hate you, I'm never coming back!* At the end of one day's frustrated foraging, his stomach still aching, he simply keeps walking instead of going back to his home-tree, where there are zero things to eat. Sometimes when he passes trees, he pulls himself up for nostalgia's sake, but mainly he walks. The rhythm soothes him and any misgivings he had are quashed. Walking is more efficient in terms of speed, of noticing danger, and of freeing up limbs for defensive action. If his absence is noticed back home, none of his family give an indication of minding—although his aunt develops a facial tick which may or may not be due to missing him.

~~~

Risk-taking is the genetic umbrella under which sits dispositions of the nomadic or grass-is-always-greener variety. (Which, in Freddie's case, is not a metaphor. Sometimes the grass really is greener.) I inherited a nomadic gene, along with a lot of Americans descended from risk-taking European immigrants. In the first third of my life I moved thirty-nine times to places where I stayed long enough to receive mail. Then I settled down. Kids were born, grew up, then grandkids came along. Pets came and went. Cats, dogs, a goat, twelve rabbits, and forty-five guinea pigs filled the pet cemetery at the bottom of the garden. Trees were planted, grew, then were chopped down because they blocked the view. Constant change, even though I stayed in the same house. I guess the nesting gene beat the nomad gene in the end.

~~~

Meanwhile, Freddie walks and walks. Loses weight and gains muscle. There's a hominid about one hundred and seventy-five miles northwest who doesn't belong to Freddie's family, but bears some resemblance in the fur department. Lots of it tufted around her legs, chest and arms, less so on her rump and around her eyes and mouth. Her nostrils and mouth are far bigger than Freddie's, and her forehead is far lower. Like him, she's a bit different from her siblings. Her spine is slightly longer and stronger than theirs. It supports her back so well she can walk upright for long periods, which she mostly chooses not to do when in company because she intuits oddness is not a good thing. She doesn't like the attention it draws. Mostly she climbs or runs with all fours, her legs swinging through her arms. It's easy and even fun sometimes, but it feels false. Aside from both being misfits, she and Freddie are as different as chalk and cheese. She loves to eat rats and the occasional fish, he still prefers fruit. She hates fleas, he can't get enough of them. Her name is Fiona and she's my grandmother times a million. Possibly yours too.

Freddie meets Fiona and without giving it a minute's thought, one moonlit night with the sound of water rushing (for he's near the sea now, near what will one day be Luanda) he mounts her. He pumps away for exactly six seconds while Fiona crouches stock still.

You could think of this as another Adam and Eve moment from which we eventually spring. Yes, we know that Kevin splitting was *it*. But the idea of originating from an androgynous thing-like-a-virus is much less appealing than this moonlit coupling. At any rate, Freddie impregnates Fiona, their two gene pools combine and voila! Their first baby has a high forehead like Freddie and a strong back like Fiona, and so does the next baby. The third has a small head and doesn't survive childhood, much to Freddie's surprise. Yes, he's still around, because monogamy is starting to be a thing—not for all hominids, but some. It increases survival chances when the father is around to protect the infants. The children of the more promiscuous members tend not to survive, much less reproduce. It would be nice to imagine some tenderness is involved. And who knows, maybe it is.

~~~

Some of Freddie's family from Hominid-ville will eventually head north too, but only the ones who master walking on two legs will make it. His aunt will succeed—it's as if Freddie left a bread crumb trail visible just to her—but 75% of the others fall to the ground and are left behind without a backward glance. Migration weeds out the weak, and the next generation is hardier.

~~~

But all that is in the future. Let's go back to the drying-out place, with toddler Freddie eating red oblong fruits.

'Freddie,' I say softly. He doesn't look at me, so I say what I used to say to my kids when they ignored me: 'Freddie, sugar-pie honey-pop. I've got something nice in my pocket for you.'

'What is it?'

'A birthday present.'

He toddles to me, arms outstretched, and I show him a small ripe banana.

'Is it food?'

I don't remind him he told me I was bananas half an hour ago.

'It is.' I peel it, place it in his hand and watch as he smashes the whole thing into his mouth.

'Thank you.'

Don't talk with your mouth full, I almost say. 'You're very welcome. Happy birthday my dear.'

I don't tell him this is the first time in evolution I have called anyone dear. It seems the closer I get to my own appearance, the more I gush.. Then he scrambles up the tree, causing a small avalanche of dead leaves to fall on me. Huge yellow crispy things, and I cheat (removing any natural object is forbidden) by tucking one into my pocket for a souvenir. All the way home, I keep my hand out of my pocket, aware I could crush it. I don't want to do that. It's been a long day, and I need to make oatmeal cookies.

Cora

1.8 million years ago
Still the place which will be Africa

Something has been going a little wonky with Earth's tectonic plates, and on a regular basis she over-balances on her axis—all of which would be fine if it didn't have such a catastrophic effect on the weather. Life with a capital L struggles, will always struggle, with changing climates, and you better put on your thermals because now half the world is covered in ice. We're a million years into the fifth major ice age since Kevin's time. The phrase *ice age* sounds so melodramatic, but ice ages come and go extremely slowly, taking tens of thousands of years from start to end, with long arid periods in between. It's like menopause in very slow motion. As a matter of fact, as previously mentioned, this particular ice age is still ebbing in the twenty-first century. Maybe Earth is wondering whether it's worth trying HRT after all, or should she just ride it out. It's weird to think we're emerging from an ice age. Global warming feels…well, warm.

I'm not alone.

'You're not in an ice age,' says a twenty-five-inch-high child who is standing on my toes, not far from where Kevin once floated before the continents came into being, broke apart, drifted and reformed.

'Yes, we are. And shush, Cora. It's not time for you yet.'

'You cannot possibly call your time an ice age. You have no idea. No. I. Dee. Ah.'

'Be patient.'

~~~

Like migration, you can't beat an ice age for weeding out the weak. But the good news is, it's killed off most of the dumb hominids. Smarty pants Cora is our case in point. She's thriving because she's not dumb.

'Can I talk now? My turn! Can I, can I, can I?'

'Oh my God. Alright. Yes, Cora.'

'Yippee!"

I've met hundreds of my pre-historic ancestors, and she's the most enthusiastic. Think of cheerleaders—thousands of them—packed into a one-year-old's body. I've been watching her from a distance for fifteen minutes, and am exhausted already. I crouch in front of her.

'I'm one of your descendants. Want to shake hands?'

Pause, while she lowers her eyes and skirts around my feet. No hand shaking occurs.

'I'm only one. I can't talk,' she says in a fake-baby tone of voice.

'Don't be silly. If one-celled Kevin can talk, you've no excuse.'

'Okay, okay, I can talk. Am I a Homo sapiens yet? Am I, am I, am I?'

'No, Cora.'

'Okay. Okay. Okay. Am I a Homo sapiens now?'

'No! Jesus, how many times do I have to tell you?'

'Okay. Just double-checking.'

'You're a Homo erectus. In fact, in Homo erectus terms, you and your family are the crème de la crème. Pretty soon you're going to dominate the world, and then you're going to turn into us.'

'I knew it! I knew it! I knew it!'

'What did you know?'

'I'm the boss!'

'You're confident, anyway. But you don't know everything. No one does.'

~~~

What we know increases but continues to be very little and includes this: We cannot create or destroy matter or energy, only change them. Nothing in our body is alive in itself. We are an assemblage of inanimate matter, down to the very tiniest electron orbiting the tiniest atom. The I that is I (and the you that is you) animates the body for a limited period. Feels at home in it. Grows it and wreaks changes on it. This would seem to give the AI-as-sentient-being argument some weight. After all, like computers we too are made from life-less particles. What makes us so sure Life can't animate what it chooses to animate? Or maybe we are the Life-creators like the God (or gods) some of us imagine created us. Quite a few people are nervous about AI's outgrowing and maybe even destroying their creators—but doesn't that also describe what Homo sapiens have done? Maybe we are an existential experiment that backfired when we began to think for ourselves and ignore the sensible commands of our creators/programmers. But these are disturbing ideas to contemplate at our stage of evolution, so let's not. We know that the I that is I ceases to be alive—but we don't know why it ceases, why our cells have a limited life span instead of perpetually renewing. We end and our body returns to being lifeless matter. It decays, becomes other things, and eventually at least part of it is likely to re-materialise as part of another life form. Human or not. To this extent, the Buddhists nail it with their reincarnation theory. We have many lives. Some of the lives will no doubt be lowly lives, like being part of a worm or a beetle. But scientists (the ones who are not Buddhists) claim this much: the inanimate material that was once you will not re-materialise as a single life form, much less a single person. No, much more likely you will be fragmented into thousands of life forms over time. Your former eyelash will contribute via decaying, then partially converting to a worm who ingests it, to the bird who eats the worm, to the cat who eats the bird, etc. Think of the nursery rhyme *There was an old woman who swallowed a fly, I don't why she swallowed the fly…perhaps she'll die.* Then she goes on to swallow a spider, a bird, a cat, a dog—all to catch the fly, who wiggled and tickled and jiggled inside her. You have been and will be many things. Parts of you may or may not be conscious of this. Probably not.

'What are you humming?' asks Cora.

'Oh, just an old song. A song for kids about things swallowing smaller things, then being swallowed themselves by bigger things. It's kind of silly.'

Pause while she screws her face up.

'Huh. So basically a life cycle song?'

'Uh. Yup.'

I'm thinking of Gate of Angels now by Penelope Fitgerald. Professor Henry Flowerdew tells his science students to never forget that energy and matter are also a part of themselves. But I'm not going to elaborate. Cora's only one.

~~~

She's the size of a normal one year old child and her parents are roughly the size of you and me. We're taller but their legs are longer. Their jaws and teeth are bigger than ours. If you shaved and dressed them, it might be disorienting how similar they are to us. But they are not Homo sapiens yet in the brain department. Who decides this, who makes these categories of species? Why we do, of course, because we are the master race and no one is as smart as us. *Ha ha!*

Cora's family are preparing to move. You can see them over there by the cave entrance, packing what looks like dried meat and fish into pouches made of skin. Someone—gender unidentifiable under all the fur—is stowing stone-headed spears and axes. They've contrived a sledge—a long tree sliced down the middle and hollowed-out—to slide over the snow. Presumably Cora will be strapped onto that with some stringy animal intestines, along with her young cousins and siblings. Bear skins may or may not be thrown over them. You and I think it's freezing here, but it's actually a lot milder than it's been for centuries. This does not signal the end of the ice age; it's a temporary respite and soon they'll be back in the deep freeze. They're taking advantage of the warmer weather to hit the road. There's been talk of a better life southward, and Cora's people are all about progress. Their nomadic streak serves them well, and there's a feeling in the air that something exciting is about to begin.

'Cora, do you know how old you are today?'

'What's old?'

'Old is…it's a way of talking about time. And time is…it's a word we use to measure…measure distance. Sort of.'

'Distance of what?'

'Well, our decay.' Quite liberating, being blunt about mortality with a kid. And I think she appreciates it. Nothing coy about Cora. Anyway, it's freezing and I'm running out of energy now. She opens her mouth and I can see by the shape it's going to be another why, so I cut her off at the pass. 'You are one year old today, and that's why I'm here.'

I stand up a bit straighter, to give my mission more gravity.

'Why?'

'Because it's an anniversary us more modern beings like to celebrate.'

'Why?'

'To mark our survival, I suppose.'

'But why do you want to do that?'

She's a typical toddler in so many ways—many whys, and I suspect half the time she just asks to be annoying. Maybe the closer I get to us, the less mature the ancestors are. It would make sense. If you only live a few days like Kevin, you've got to grow up fast.

'Close your eyes and hold out your hands.'

She does this. They are pink and hairless on the palm, and I place a small present in her hands. She rips off the paper and makes a gratifying noise, which I roughly interpret as *Wow*.

'What is it?' She's shovelling it in her mouth, crumbs everywhere.

'It's a mince pie.' I know. Not ideal. I wanted to give her waterproof gloves or a hot water bottle, but those kinds of gifts are not allowed. Anything that has a brand or label is also verboten. Imagine the confusion if an archaeologist today excavated a prehistoric plastic Tesco bag. And besides, we have a surplus of mince pies. Every Christmas I rediscover how much I love them, and then two weeks later rediscover they are too rich.

'More! More! More!'

'All gone. Sorry!' I show her my empty hands and tears well up in her eyes.

~~~

Cora and her family might live in caves, but they are not cave people. Not remotely the Flintstones or cave men from 2001 a Space Odyssey. We're in a place that looks like the inside of my freezer when it's long due a defrost, but no one looks starving or even particularly cold. There's some markers over there, stones in piles on the snowy ground. At first I assume some field has been cleared for ploughing, but there's only ice and snow here. The stones cover dead members of the community. Apparently they're mourned when they die—probably not extensively because staying alive is so time consuming, but still. Their deaths are marked with respect and maybe even ritual. One of the buried is related to Cora, and therefore me.

'Do you miss your grandmother?' I ask her, taking a guess.

'What's a grandmother?'

'The old woman who loved you. Your mom or dad's mother.'
'Oh her. She's not gone.'
I indicate the pile of stones.
'That's just her body! She's not in her body anymore, silly.'
'Well. That's good news,' I say. What else can I say?
'Why, why, why is it good news?' She's stepping on my toes again as she hops around, with her perennial whys.
'Because being properly dead would be...huh...less fun?'
'Why? Why? Why?'
Oh, I give up.

~~~

One of the reasons her family has not died out is because they became hunter-gatherers about a million years ago. Their cousins who migrated to Australia stuck to vegetarianism, which may or may not be why they are no longer around. Not many berries to eat in an ice age. Plenty of freezing and exhausted animals to eat. Cora's family eat a lot of meat and fish, and this partially accounts for their larger brains, which in turn might have led them to solve the mystery of fire. No one knows when people began to use fire, and looking around I see no evidence, but it's hard to imagine surviving in this place with nothing to cook on, to keep warm by, maybe even to scare predators away.

And socially, it's obvious these ice age survivors have learned to cooperate. To live in family groups. They probably don't bother much with monogamy, but the mother and child bond is definitely strong—hence the high survival rate of infants to age of reproduction. Good mothering is a major key to survival.

'Cora, do you know who I am?'
'A weirdo.'
'Aside from that.'
'Who are you?'
'I am your descendant. Or one of them. There's millions.

Maybe billions.'

'Okay. Huh. I've got to go now. Here comes Mum.'

I look up, and she's right. Her mother is walking over to us, and I recognise her expression. She's cranky and she means business. I whisper goodbye to Cora just before she's scooped up and tucked under one arm. The mother does not see me, or chooses to ignore me. Maybe she thinks I'm the imaginary playmate Cora's been going on about.

Early Homo Sapiens

Elizabeth, Ned and Eve

150,000 years ago & 175,000 years ago
Near where the Swiss/Italian border will be

The evolution clock is a minute to midnight, and Homo sapiens are here at last! Here we are with one-year-old Elizabeth. Who is not and never will be a queen like her future namesake in England, although in thirteen years Ned will treat her like a queen so he can have his wicked way with her. Which will take just under forty seconds and from which I will eventually emerge. She won't think of herself as a snob, but as soon as it's over, she'll realise her misjudgement and cringe. Ned is someone about whom she'll never say: *Guess who's coming to dinner?* Because Ned is a neanderthal and she is a Homo sapiens.

But they are both hominids, of course, so it's not like a fish mating with a giraffe. *Still.* Ned may be attractive in a brutish blood-pounding way, and it's quite sexy the way he mostly doesn't say anything, his dark eyes smouldering—but she'd never respect him in the long run. Nor would her family accept him. Within a day, she'll think of it as an ill-judged fling, one she won't regret as long as she never has to see him again. Personally, I thank God for Ned.

'Excuse me. Excuse me.'

Oh my God, she can hear me. And like cheerleader Cora, she's making the first move. I'm not ready for this. Who's in control here, dammit?

'Excuse me, I said. Can I ask you a personal question? Are you, by any chance, watching my future?'

She's got a very soft but clear voice, with a slight Yorkshire accent which is bizarre given where we are and the fact Yorkshire will not exist for thousands of years. An average-sized one-year-old girl with reddish brown hair, brown eyes and freckles, she looks at me with a sweet seriousness—but also gently accusatory. Fair enough, I think.

'I said, are you watching my future?' Very slowly this time, as if she suspects I am slow.

'Who, me? No! No, no, no!'

'No?'

'Well, not intentionally. I mean, I'm not spying, *per se*.' Another lie.

'Oh. Okay. What am I doing in my future?'

'Oh, nothing. Just, uh, playing a fun game in the woods. Like tig.'

'What's tig?'

'You'll find out.'

'Is it where one person chases everyone, and then whoever he touches gets to be the next person who has to chase everyone?'

'Yeah. It is, actually,' I manage to croak.

'Oh, we call that boog here. It's an old game. Everyone knows it.'

'Well, I'll be darned!'

Then a pause while we both study each other. I think she's as curious as I am.

'Okay. Right,' she says slowly. 'And can you please remind me, what exactly are you doing here?'

'I'm here because it's your first birthday. I'm one of your millions of descendants.'

'Huh. Wait a minute.' She presses her lips together in concentration and I wait.

'You mean one day I'm going to have a baby, who will have a baby, who will have a baby, and then eventually one of the babies will be you?'

'Correct! You were born with the beginnings of me inside you. Just like all my potential descendants were tucked inside me when I was born. It's an egg thing.'

Oh, it is such a delight talking to someone as curious as me. I'm so excited, I get carried away and tell her how much I'm enjoying getting a glimpse of another of my forebearers.

'I don't often get to see both lines.'

'What are you talking about?' she asks.

'Ned!' *The Neanderthal*, I almost blurt.

'Who?'

'Oh, shit. Sorry.'

I've definitely broken the law now. I glance around quickly for camouflaged cameras or mics.

'Who's Ned?'

'Just someone nice you're going to meet one day. Don't worry about it. In fact, forget I mentioned it at all.'

'Mentioned what? Did you bring me a birthday present?'

I'm beginning to think this whole story is in my head, it's going so well.

~~~

It's not always easy tracing your family line, and I can't find an earlier Homo sapiens grandparent. I contain more of her than any other previous ancestor, as the further back I go, the smaller the fraction of genetic input. That is to say, her DNA is inside some strands of my DNA, which reside inside the nucleus of my genes, and these exist in every part of me including toenails and nose hairs, but not my eggs. They have only one copy of each chromosome, rather than the two that exist in all other cells. This is a permanent situation. As long as my offspring keep having offspring who keep having offspring, my version of Elizabeth is immortal.

Of Kevin, I contain the scrapiest scrap imaginable, and so do you. Strange, but biologically a fact—every life form is related to every life form past and present. Some animals are more similar to us than others. For instance, we share 99.99% of our DNA with other people, with only .01% to make us an individual. Eye and hair colour, curliness of hair, bushiness of eyebrows, height, sense of humour, freckles, double jointed thumbs, propensity to obesity or alcoholism or depression or to producing great works of genius, preference for autumn over spring, for avocados over hamburgers—all of these traits and more, squeezed into that .01%. As for chimpanzees and bonobos and gorillas, well! Almost no difference between us—about 1%. Have we evolved more successfully? It's debatable. As long as a species is alive, it's successful. Ants are successful. We can't even look down on fleas. And it's not just animals we're related to. We share 31% of our genes with yeast, which is only one cell big and replicates every ninety minutes. Weirder, we are 50% genetically the same as bananas.

Think about it. *Bananas.*

Strangely, we still shy from lumping ourselves in with the animal kingdom, much less fruit and vegetables. We're vain. There is glorious us, and there are animals and plants. Dogs we throw sticks for, or cows we milk, or food we grow to eat. The more exotic animals, we let David Attenborough tell us about on television or we visit in a zoo—which is essentially a prison for animals who've done nothing wrong aside from being born not human. We're so convinced of our superiority, of our prior and God-given rights, this scenario is not even a subject for discussion. Unless you find yourself in a group of vegans or vegetarians or perhaps Buddhists. You'll recognise them by their youthful glow and slender physiques.

~~~

But let's talk about Ned the Neanderthal, my other ancestor. He's part of the equation and deserves more than the role he's usually given. Were his eyes really smouldering for pretty Elizabeth, or was he just obeying his genetic insistence to keep his race going by any means available—unaware, as we mostly and luckily are, that his species is doomed? Maybe we all rationalise behaviour and choices, while at heart we're just animals obeying impulses imbedded in our DNA. Instincts that spring from our genes, reside in our subconscious, and mostly do what they want—no matter how much we think we run the show. It's anarchy! Or maybe just mutiny. Either way, this theory can take the sting out of shame when you've botched something. You're programmed to botch it.

What Life needs to keep going, Life gets. Sometimes it's pretty, but not often. In about 120,000 years Neanderthals will be extinct, but their DNA will be embedded in us forever. Or, to be more accurate, until we become extinct—and not even then, if we've evolved into another life form. Maybe there should be a word for extinct life forms that live on in other life forms. But wait—doesn't that describe absolutely all life forms since Kevin? None of us has magically appeared out of nowhere. We're alive, and therefore need to give credit to now-defunct species. The words *new species* have less clarity considered this way. Can there be such a thing, and if so, what defines it? Scientists, as usual, have an array of contradictory answers.

~~~

Elizabeth is not my only ancestor to fall for Neanderthal charm. I have, according to a DNA test based on my saliva, more Neanderthal variants than 94% of their customers, which number over nine million. This has nudged me to think about the Neanderthal in a more complimentary light, not the oafish cave man grunting and wielding a big stick. It might also explain my red hair, for Neanderthal DNA shows that some were red-haired—and by the way, you can blame red hair on chromosome 16 and the MC1R gene mutation. These days, only 1%–2% people worldwide have red hair. This rises to about 10% in Ireland, which might mean the precursors of the Celts, like Elizabeth, did some inter-species dating. Or it might not. No one knows, but I reckon my hair colour originated right here, with Elizabeth's ill-judged fling.

~~~

They might not seem the most successful species on Earth, given their extinction, but I'm fond of Neanderthals now. Who doesn't root for the underdog? They remind me of the Cal Bears, the college football team our family always supported despite (or because of) the fact they lost 99% of the time. We often call Neanderthals *cave men*, though in general they do not live in caves. There are simply not enough caves. Their bones are mostly found in caves for the obvious reason that modern man cannot build cities and roads on top of caves. Which means caves are one of the few places left undisturbed; hence their popularity with archaeologists. The truth is, we walk on the remains of our ancestors when we go to the store, when we walk from our kitchen to the living room, when we walk the dog in the park. Not as intact as bones found in caves, but everyone who ever lived is still here in some form, sometimes right under our feet. Or if it's windy, up our nostrils.

In another sense, you walk through your ancestors every day and aren't aware because unless (like me) you're lucky, you have no concrete access to past times. What if: There they all are, going about their own unaware days, foraging, sleeping, farting, having sex, splitting, spitting, dying, coughing, what have you...and here we all are, doing exactly the same things in the same places, and not even noticing when we walk right through them? What if there is no *now* that is definitive for everyone? Do you ever feel suddenly tired, as if the air has become thicker or gravity is exerting more force? Or notice when an atmosphere turns inexplicably unhappy or tense or giddy? Maybe you're overlapping with a dense crowd of past beings, or you've stumbled into a pocket of emotional intensity. Maybe someone's just died or won a version of the lottery. We emit measurable amounts of electricity all our lives, all life forms do. If we're sitting around daydreaming, it's about 100 watts. If we've just won a race or fallen in love, it's at least 2000 watts. Just because someone lived five centuries ago, maybe that's no reason the ions of their sudden dip into melancholy aren't burning through to our today. Maybe we exist in a timeless soup of everyone who ever lived, because in a sense they are still living. Or maybe Time is chronological after all. Life happens as we perceive it and now means NOW, and nobody from the past has an existence in the present. That's much easier to swallow, though nobody knows these things for sure.

~~~

But back to Ned the Neanderthal. Most of his species live in small population groups in Europe (but not north of Belgium) and southwest Asia. Italy has one of the biggest clusters—another reason to like Italians. Aside from owing Ned and dozens of other Neanderthals my life, their genes theoretically help me fight viruses and conceive babies. My Ned-red hair also gives me skin that is so low in melanin, it's hyper-efficient at absorbing vitamin D in very low light conditions—such as winter in the Scottish Highlands. Not so good when I have a week in Costa Brava, however. Low melanin can equal skin cancer.

The Neanderthals have a good thing going for a long time—

they last 200,000 years longer than we've existed already. But one day, in about 128,000 years (28,000 years ago from us now) there'll be just one Neanderthal left. A red-haired female called Sadie who calls a Gibraltar coastal cave *home*. She's outlived not just her family and friends, but her species. She will be, in the official terminology, an Endling. She won't know she's an Endling—no one ever does—but she'll feel increasingly listless even when her stomach is full, and she won't know what's going on. She'll pace a lot, scratch her itches so often her skin will bleed. She'll make friends with a bonobo, but it won't go well because the bonobo's family keep biting her. Then one day, after she loses her appetite and stops peeing, she'll go to the deepest part of the cave, curl up and stop breathing.

Why have all the Neanderthals gone?

It could be that Homo sapiens flee inhabitable places like Africa to places with more resources and a better climate, like Europe. And then we brutally colonise. We grab land, hunt predators into extinction, and wipe out the Neanderthals.

Or, more flatteringly, Homo sapiens are not the bad guys. Homo sapiens means wise man, after all. Maybe we delay moving into Europe until the scary Neanderthals are no longer a threat. The Neanderthals might have dwindled due to excessive inbreeding. They live in small not-very-nomadic family groups, after all, and over the centuries weaknesses might have been reinforced. Weaker immune systems. A lethal virus might be the culprit. Maybe I am too inclined to loathe my own species. It's very possible. I blame it on my Catholic upbringing. If something's gone wrong, anywhere to anyone, I'm guilty. I'm a Homo who's not so sapiens.

~~~

Meanwhile, what's happening to Elizabeth's non-extinct hominid cousins? The ones we did not descend from but share ancestors with. Like us, the apes, bonobos and gorillas are still evolving. They may never dominate Earth, but in their own way they're doing just dandy. Or are they? I find it very uncomfortable, looking at those close-up photographs of mother gorillas in zoos, holding their young. Sometimes kissing them. Their poses are so familiar, so like us—and their eyes look intelligent and sometimes dead. As if they're depressed but plotting something. Think of Planet of the Apes. It's only a poorly situated larynx that prevents them talking like we do. Though of course, gorillas talk to each other all the time. We just don't understand it.

~~~

But let's get back to little Elizabeth, my suspicious and eloquent Homo sapiens great grandmother times a million who never gives a thought about her origins or the nature of the universe. She has a famous hominid grandmother (times a lot) who came from east Africa a hundred thousand years ago. Evolution scientists call her our mitochondrial Eve because she is the most recent hominid ancestor common to us all. We are all descended from her in an unbroken line through our mothers and our mother's mothers, ad infinitum. In other words, everyone alive today has an ancestral line that converges on this one woman. She's what we have in common—in addition to, of course, Kevin. Amazingly, none of her descendants have the exact same gene variation—which might explain our success (so far) as a species. I wanted to visit her on her first birthday, but boy oh boy is she hard to find. Elizabeth is here in front of me right now, so why fret about missing the celebrity interview?

'Do you know your grandmother times a lot was super famous?'

'Yeah, right.'

Her accent is morphing into modern American adolescent

now. I can't account for this, any more than I can account for the way a one-year-old communicates with me at all. I hear Elizabeth's lucid voice, even though all that's coming out of her mouth is *goo goo ga ga*. It's uncanny, and it happens with each birthday visit. Thinking this, then thinking about things I have to do before the grandkids arrive, like bake peanut butter cookies and set out the paints, I trip over a root, and…*shit, shit, shit.*

I'm so sorry. I seem to have lost Elizabeth, and we didn't even get a chance to say goodbye. Tuning into ancestors is hard. Some days I'm good at it, other days not so good. But oh my, who on earth is in front of me now? If it's who I think it is, how ironic. After all those failed attempts, I've stumbled upon her accidentally. Elizabeth's grandmother times a lot. I'm whispering for a David Attenborough reason: I don't want her to notice me. This visit is too unplanned, and I'm dumbstruck. It's like suddenly finding yourself alone in a lift with your favourite actor. You thought you'd give anything to meet them, but now they're here, all you want to do is throw up.

Here. She. Is.

Eve.

I know this place is hardly Eden, but believe me—this is Eve, and she's pivotal. From here on in, Homo sapiens are inevitable. (Again). I'm going to give myself a few minutes to calm down. Take deep breaths. Think of things I've done right. Of making perfect cinnamon toast on a snowy day for my kids in front of the fire. Of hitchhiking all the way from Alaska to San Francisco with one lift. Of choosing the father of my children. There. That's better.

We should celebrate Eve's first birthday too, which (of course) is today.

Look at her.

Alright, not as cute as Elizabeth, her granddaughter times a lot, but now you know who she is, the role she plays in your life, maybe you can imagine she displays a sort of charisma. Her eyes are close set and tiny, and I detect an assessing calm intelligence. A light in her iris. As if she's thinking:

Do I want this fruit right now? Maybe I should hide it someplace where it can dry out, and then later when it's freezing and I'm hungry I'll have something to eat.

And intelligence is very attractive in itself, don't you think? Maybe a fling with a Neanderthal is fun, but we evolve wanting to breed with intelligent beings, even if they're obese or ugly or selfish. Maybe intelligence is the most powerful survival tool of all. Or perhaps it's more gender-based. If we're female, we're programmed to desire a tall intelligent healthy mate, but even in Elizabeth's time it seems to be the curse of hyper intelligent females that not many potential mates approach them. Maybe boys want the pretty girls who don't make them feel like losers but still give them babies. Maybe some smart girls of a certain age settle for anyone who can give them babies. Like Polly in the sea off Goa, sighing with relief and thinking: *At long last!*

~~~

'Happy birthday, Eve,' I whisper nervously. I wonder if my hair looks alright.

'Who the hell are you?'

'I'm your eventual outcome. One of them. In fact, every Homo sapiens alive is your outcome.'

'What *the fuck* are you talking about?'

None of my ancestors so far have been crude. Is it her culture, and I shouldn't read too much into it? She's scary.

'Ha ha happy ba ba ba birthday, Eve!' I stutter.

'Can I eat you?'

'No.

'Get out of my face, then.'

'But you're a herbivore,' I whisper, shocked. 'Aside from the occasional bug.'

'You think?'

I close my eyes, count to three, and when I open them—Eve is gone. Whew! And by a miracle, we're back with Elizabeth. She's looking at me accusingly again.

'Sorry, Elizabeth. I got distracted.' I'm even more struck by how biddable she is, now I've met Eve.

'By my future again?'

'No!' Such a relief to tell the truth.

'So,' she says, folding her arms. 'Where *is* my birthday present?'

'Right here!' I pass her a small wrapped gift and she tears it open. Out falls a plum.

'Oh! Oh! I've always wanted one of these!' She dances on her toes.

'Do you know what it is?'

'Nope!'

Then she eats it, spits out the stone and skips away from me, suddenly more like a four-year-old than a one old. But that's the past for you. No one acts their age.

Hans

70,000 years ago
What will be Germany

History, it seems to me, is largely a series of accidents. Sure, there's intentions and plans, but overall—randomness. Movements of beings (from protozoa to hominids) towards and away from each other, often not for the reasons they tell themselves. The urge to connect, aligned with evolving awareness that true connection is impossible. We are all locked into our own little worlds, right? On and on, no essential change. At the same time, continual change.

Something big has happened since Elizabeth met Ned the Neanderthal. *Homo sapiens are different.* We've begun moving rapidly across continents. Indeed, across the planet. We're learning to build boats. To create settlements thousands of miles from our grandparents. We're not interbreeding much anymore with the Neanderthals and a new kind of cognition is emerging. Human consciousness if you like. In particular, creative imagination is growing. Light bulb moments are regular occurrences, hundreds of centuries before the invention of electric light. Art and music were born long ago, but now both have become sophisticated. (Have a look at the ochre paintings in the Blombos caves in South Africa, if you don't believe me.) Why now? Maybe a genetic mutation has altered our cerebral wiring. Or maybe we've just become less preoccupied with mere survival, now we're starting to get the hang of growing food. Nobody knows for sure, but by the time of Hans's first birthday, some of us are able to convey not just information but imagined worlds. Made-up stories from 70,000 years ago! Even if they were just telling lies in order to procure the best piece of bison, that's pretty impressive. Does this mean we're even gossiping? Yes. *God, yes.* What's more civilised than gossip? Some anthropologists believe gossip evolved as a kind of social grooming, to replace literal grooming. According to contemporary historian Yusaf Noah Harari, this is the cognitive revolution from which our current society springs. Right here, right now. Well, so to speak. Margaret Mead would beg to differ. She'd say civilisation began when the first femur bone was allowed to heal, thereby proving that someone cared enough to look after someone else. That compassion, not intelligence, is the springboard to society.

~~~

We're in what will become Germany. Hans is that boy sitting next to that old man wearing what looks like a badger on his head. Ignore him. I tried to make eye contact earlier and he spat at me. The Homo sapiens who made Hans mainly came from the southern tip of Europe and northern Africa. His mother's mother came all the way from the same hominid town in Africa that Freddie (my grandfather times a thousand) left one long-ago hungry day. The one who, instead of going home as usual, just kept going north. Since his exodus, migration has become a popular group activity. Han's father's grandmother is unusual (which in evolutionary terms, is a good thing, remember?) in that she travelled south, from what will be Holland. Originally she was travelling with a group, but they got lost in a snowstorm. Some died and were eaten by others, and in the end she was alone. *Thank God she made it.* Another close call to me not being me. Anyway, I am related to Hans on both sides. That makes me feel more connected to this forebearer. Also, he's pretty cute. His longish hair is light brown—which is as close as you can get to blond currently, given blond hair has not appeared on a single Homo sapiens scalp yet. The blond gene will come to Europe in about 50,000 years courtesy of the western steppe herders from southern Siberia where the lack of sunlight will lead to a mutation in pigmentation DNA. In most European countries, aside from Scandinavian ones, blond hair will remain rare enough to be valued highly, and blond females will have an easy time attracting mates. Hans has never seen blond hair, and would be gobsmacked to know he'll have hundreds of blond descendants. Makes me wonder about the appearance of my own distant descendants. Anything is possible. Maybe with global warming pale skin will exit the gene pool, and we will all be black skinned. Again.

~~~

'Hans,' I say with reverence. Hands together, like I'm praying.
No response.

'Hans,' I say again louder, because I'm not sure he hears me.

There is such seriousness to his little face, I mustn't smile too readily. *No gushing*, I tell myself. I adjust my tone to be casual and deadpan. Almost monotone. He looks at me suddenly and intently. I flinch because it's like headlights are now shining on me. There's thoughtfulness in his eyes, and maturity. More mature than my kids at one, anyway. Is it unnerving? A little.

'Happy birthday, Hans.'

I wait for an equally sombre response. I will refrain from hugging this child, although he is eminently huggable. He stares at me for a full minute.

'Crapolino!' he finally says in an Alabama accent, bursting out with an infectious laugh bordering on irritating. Like a machine gun. 'Am I finally one? Jesus Christ almighty, I thought this day was never going to come.'

'Huh,' I say, while I marshal my thoughts.

'Yackadackadoo!' I answer, trying to be as silly as I can to establish a rapport. 'The day is here, your majesty. You are now a…a baby man.'

'Make my day, why don't you. Got any cake?' It's still an American accent, but East coast now, mixed with a Latino vibe. I don't even try to account for this.

'Oh goodness me. I forgot the birthday cake!' I throw my hands up in mock despair.

Hans's face falls, tears form. I count to three, then pull a cake out from my pocket (which is huge, the jacket having been purchased with these visits in mind). It's just a Dundee cake, shop-bought, but his eyes light up like it's a gourmet gateau. Now I wish I'd invested in a proper chocolate cake with cream. Better yet, I could've baked one of my honey cakes with sesame seed sides and butter icing on the top. Damn! Already, I'm thinking I might need to return with a proper cake. This boy would so obviously appreciate it.

Hans wriggles away from old man badger-hat, and bum-shuffles towards me. He's not wearing a nappy, but his long overshirt protects his nether regions. I suppose he'll be trained like a puppy. Lifted outside when he starts to poop. I look down at the stone floor and spot several puddles which could be baby pee. No one seems to mind. And by no one, I mean badger man and the three older children who have just wandered in with some kind of pet rodent following. Rat-like, but twice the size and short-tailed. Maybe they all smell the cake. There's a sense of barely contained but cosy chaos here. My kind of place, I think. Kids, pee puddles, rodents, cake.

I open the wrapper and carefully break off some of the cake. There's a stack of pottery plates and bowls on a table nearby—three of which are currently on display in a museum, part of an exhibit called Linear Culture. An orderly and well-equipped dining table means this could be a civilised birthday party. Good. But Dundee cakes are covered in whole almonds, and I suddenly wonder if one-year-olds can eat things like this. Safe or stupid? But Hans snatches it from me before I can decide. He shoves it in his mouth. Swallows it immediately, then grabs the rest from my hand. He's so fast, I can't stop him. The others are scrabbling about on the floor for crumbs. The rat-thing is scampering over their shoulders, nibbling the crumbs from their cheeks. It's making purring sounds, but not like a cat, more like a car idling. No one seems to mind anything. *This is not going how I thought it would go.*

'Can you at least chew properly? You'll choke,' I can't help saying. I'm using my bossy voice now. Back to being me. I like to think I'm laid back, but my tidier self often leaps out at times like this.

'Fuck you,' Hans mumbles. Then he opens his mouth and smiles broadly. Only three teeth. Two lower front, one upper front. Chewing might not have been possible, I think.

'Mind your language!' I say automatically.

'*Fank you*?' he repeats innocently, this time with an inflection at the end.

'Oh, okay. You're welcome.' I'm suspicious. He definitely said crapolino earlier. He might be messing with me.

~~~

But it's hard to be hard on a kid who's about to lose it all. Yusaf Noah Harari's cognitive revolution is about to take a huge hit. A volcano in Indonesia just erupted. Mount Toba, on Lake Toba outside Sumatra in Indonesia. It's the biggest volcanic eruption in Earth's history before or since Hans—though this will be hotly debated, like most prehistoric events. None of the arguments will affect Hans because the ash clouds are already coming this way. The whole planet is going to slip into an ash winter that will last about eleven years. Of course it's still hot in Sumatra (where everyone is buried in lava), but the ash clouds will soon lower the global temperature by three degrees. Like Narnia, during the reign of the ice queen. Germany-to-be will lose its blue skies for so long, people will think the gods are punishing them. That the apocalypse has come. Hans's community has 554 people, only four of whom will be here next year. Not everyone will starve due to crop failure or freeze to death. Quite a few will die violently in the anarchic scramble for survival. So many bodies, no one will bother to count, much less bury them. Everywhere on Earth, the same decimation of plant and animal life. Within two years, there will only be 7,000 or so breeding pairs of humans left globally. *Breeding pairs* sounds like the kind of phrase you'd apply to a protected endangered species or dairy cows. We tend to forget we're animals too, programmed to replicate our DNA. As individuals, we're useless in species-survival terms. Bottom line: If we want to count, we have to make babies, or at least assist in the caring for babies. Same as pandas, same as honey bees, same as sabre-toothed tigers. Small numbers of breeding pairs causes a genetic bottleneck. The huge variety of genetic traits accumulated up until now will be mainly lost. From here on in, everyone will be a product of these 7,000 or so breeding pairs. We are all ash survivors.

In particular, I am here because Hans will make it through the ash years. He won't be a great specimen because malnutrition will stunt his growth and his personality will cease to be ebullient. (Ebullience is a luxury trait, for when there is enough to eat. Not a biological imperative.) His virility will remain intact long enough to impregnate a girl called Roxane. It happens that Roxane will be the only fertile girl left in a radius of 330 miles, so Han's only child (oh, the fragility of my link!) will have lots of half siblings. Twenty-four, to be precise, who within twenty-five years will produce 162 people. She works her butt off. *Thankyou Roxane.*

~~~

But that is the future, and one-year-old Hans is now demanding I empty out my other pocket.

'I have no more cake,' I say. I turn my pocket inside out to prove it. Some dog poop bags float to the floor, and I scoop them back into my hand as required by law.

His tears well up again—are they genuine or manipulative? Is this the birth of crocodile tears? He's so cute when he's sad.

'Close your eyes and hold out your hands.'

He complies and I place a tiny gift-wrapped box in his hands, which are pink and surprisingly clean. His nails are short and also clean. Someone looks after my great grandfather times a hundred thousand, that much is clear. I can't see any parent-type people, but maybe they work.

'Alright Hans, open your eyes.'

He squeals and rips the paper. It's a Winnie the Pooh, and I immediately see this is inappropriate. What made me do it? Am I becoming blasé? Irresponsible as those American tourists I despised in my early travelling days, filling pristine Aegean beaches with their loud crassness and litter? Then Pooh is gone because the rat thing has flown out the window with it in its mouth.

That solves that problem I think, but then Hans starts howling. If he could stand up, he'd be stomping his little feet.

'Wah, wah, wah!'

'Jiminy cricket,' I say, but he doesn't hear me.

'Hey, hey,' I say a bit louder, in a comforting way. 'Look! Look over there, is that a giant giraffe?'

He looks briefly, shuts up, but isn't distracted long and screams again. Maybe I should have chosen an animal he'd recognise.

'Wah, wah, wah!'

He cries exactly like my grandson did yesterday in Tesco when he realised he'd dropped his Tigger somewhere in the carpark and I refused to go back out in the pouring rain to look for it. That same mournful unignorable sound. Evidently some things don't change much. I can hardly bear it. I sigh, as I sighed yesterday, and leave the hovel to search for the rat-like thing and prise Pooh from its teeth. Just like I eventually surrendered yesterday. Abandoned my trolley full of food, and hand in hand with my hiccupping grandson, went back out to save Tigger from the tyres of moving cars. Two biological imperatives at work here. The ability to produce a noise that makes any other action impossible but the one you want performed—i.e. the grownup dropping everything and doing your bidding. Surely this alone has led to infant survival rates improving. And the knack of knowing where lost items are, like where did I bury those dried beans and meat? Oh yeah, under the stone by the lopsided tree. This too must account for at least a 5% increased chance of survival. In case you're wondering, I was successful in both cases. Tigger was just where he'd been dropped, face down in a muddy puddle. He needed to go straight into the Tesco toilets to be washed clean and blow dried under the hand drier. Winnie the Pooh was not so lucky. By the time I wrested it away from the rat thing, after chasing it for ten minutes up a gravelly hill, Pooh only had one limb left. Hans didn't mind, amazingly. Maybe possessing two eyes, two ears and a mouth is sufficient to cancel out any number of bear amputations. Hans squeezed his eyes in apparent ecstasy and hugged that bear so tightly, I would've cried—if I'd been an easy crier. Which I am not, much to my dismay and bewilderment. Why was I born to be eternally amateur at crying? No idea, but I'm pretty good at finding things. Though I don't always succeed. People can be slippery, and single gloves are just plain sneaky.

Norma

10,000 years ago—8000 BC
What will be Heathrow

We've come out of the Toba volcanic catastrophe by the skin of our teeth, and in any case, it's an inter-glacial period and Earth is warming up again. Yusaf Noah Harari's so-called cognitive revolution is back on! Agriculture is beginning to be a real thing. Not only is food grown to eat, it's also grown to store for eating in winter. Cooking food—killing toxins, improving taste—is pivotal to becoming modern humans. Hunters and gatherers and growers are settling down together for the first time. After a while, they even talk.

'Dude, try some of this.'

'What is it?'

'Tree root.'

'No thanks.'

'Jeez, you're a fussy eater.'

'Here. You should get your teeth into this.'

'Do you mind? You're getting that, that…blood on my shoes! What's it from?'

'Gizzard of bison. Killed this morning.'

'Definitely no thanks. Though maybe I could use the bone in my vegetable soup? I mean, if you don't mind. After you finish, just toss it in the pot.'

'Huh. Weird. Bone into water. Okay, I'll try it. Besides, I am *not* a fussy eater. I just have certain likes and dislikes.'

Alcohol is being made from fermenting fruit, which might help with these kinds of occasions. Sails are made and used, making sea travel less dependent on rowers. And everywhere—jungles, deserts, mountains, valleys—babies are increasingly living long enough to make more babies. Life is not perfect—populations living in proximity means the emergence of exploitive hierarchies and the easy spread of disease. But overall, it's looking good for humans here on Earth. I feel like a cheerleader with my rah-rah skirt and poms-poms. *Hooray for mankind!* We don't have an anthem yet, but we really should.

~~~

As I write this, the evolution clock is ticking towards midnight (and me). It's an early November morning and we're in the place which will become Heathrow airport. To be precise, in what will be the busy network of roads around Terminal Five. Buses and cars will whoosh through where I'm standing and under my feet will be sewers, water pipes and electricity cables. Lots of packed suitcases and high hopes and cranky toddlers and snappy spouses and self-important businessmen, plus a fair smattering of sleepy British Air employees—all converging on the terminal carpark. But right now, there's just Norma's house, which is half dug out of the ground and half made of wood. It's set against a huge boulder which forms the back wall.

Come on in and unzip your down jacket. Even though it's November and Britain, it's warmer here than in our time. In about 4000 years, the Earth will dip into a very cold snap which will last about 150 years. Long enough to be another source of Narnia-type legends. Then it will go back to being a mild inter-glacial period before heading back into permanent-seeming winter. The current ice age (yes, yes—the same one we are still in right now. Norma's ice age is our ice age too) is ongoing, but it's fickle. Sometimes not very ice age-y. Technically it's the Early Neolithic period, but Norma's parents just call it Modern Life—when they call it anything. What will our descendants call our period of history? I'm guessing not the Age of Wisdom and Tolerance. Probably something like the Early Plastic Period. Or The Tech Times. To be accurate, Norma's time is also the very tail end of the Stone Age. It began over two billion years ago and will end in about two thousand years, when humans move into the Bronze Age. And of course, this only describes time in Britain. Over in south-east Asia or down in Antarctica, they'll have their own timetable for such things, with their own names for eras. It's easy to forget every place has its own show going, and our perceptions mainly apply just to us—which is, of course, a skewed view. Even so, it tickles me, our need to categorise passages of time and label them. It makes me think of spring cleaning, that cathartic urge to put things where they belong and throw unneeded things out. To make sense of our houses, and our pasts. Also, because I am essentially shallow and not a scientist, phrases like Bronze Age make me think of fake tan products. And blond beach boys. And beaches smelling of coconut sunscreen. I have to rein myself in. I'm standing in the place where I will one day board 747s in Terminal Five, after six Wagamama's king oyster mushroom skewers and a glass of white wine. By age seventy, I will have ping-ponged fifty times between California and Scotland, and still be trying to decide which country I should settle in.

~~~

Back to Norma.

It's pretty dark—can you see her? Her mother is nursing her over there by the dying fire. The sound of sucking identical to the sound babies make now, of course. Over this is the singing from her mother, a low soft crooning as she rocks back and forth. There are no separate words that I can distinguish, just a single clear tone a bit like an oboe. How uncanny that the human larynx has evolved in a way that allows us to communicate and make music. Think of it—our bodies are a million amazing things, and they are also musical instruments. The song winds slowly up and down soporifically. Is this a lullaby? Or maybe a nursing song? Then the singing stops, and the gentle rocking.

'Happy birthday, Norma!' I whisper. 'Guess what—I'm your granddaughter times, well *a lot*. It's a pleasure to meet you.'

She pulls away from the nipple with a loud popping sound, and says: 'Goo goo ga ga.'

'Exactly my sentiments.' Luckily I speak Baby, although until now that hasn't been necessary.

'Wah! Goo goo.'

'Goo goo wacka dacka doo!' I say, meaning I like her too.

Her mother tries to wriggle her back into suck position, but she won't have it. She stares right at me and babbles like we're old friends. Sadly, she doesn't seem capable of much in the way of conversation. Maybe it's because she's a more advanced Homo sapiens, which means she's dumber longer than mammal babies who need to be able to feed themselves from birth. Or maybe she's just slow. Not all my ancestors can be geniuses.

'Babababo. Juuuaap?'

'No, I'm not your only descendent,' I tell her slowly in Baby. 'Look behind me. They're all obliged to you for their lives. Millions of human beings.'

She peers over my shoulder, and her eyes widen. Yikes! I only meant it as a figure of speech, but maybe my fellow descendants are all hovering behind me. A disturbing but pleasant thought. But Norma is still looking expectantly over my shoulder and I am trying to be mature, so I summon the distant sound of cheering and applause. Like Wembley stadium from a mile away.

'See? You're popular. Very.'

She smiles at them (*Not me! Wah!*) then goes back to sucking. She's wrapped in a leather garment, which might be mammoth skin and probably sewn with needles made of bone and thread made of sinews. How was the eye of the needle made? By hammering the tip of a sharp stone into a narrow bone, I guess. Her mother's face is hidden in the shadows, but her arms are visible from the elbow down—pale but strong looking. If there's hair on her arms, I can't see it. If she had a haircut and some decent clothes, she'd not look out of place meandering around Poundstretcher.

~~~

The main industry in Norma-town seems to be stonework—but much more advanced than in Elizabeth's and Hans' time. Over there on that shelf is a neat row of tools. Bone needles, stone axes and chisels, grinding stones, all lined up in order of size and use. You don't get to survive an ice age and a decade of ash clouds if you're disorganised. Or a poor tool maker or fire maker, come to that. Over by the back wall in the shadows, there's a tidy pile of fuel to burn. Branches, mostly, and dead leaves and twigs and animal fat scooped into what looks like a large human skull. The fat accelerates fire. All of these things are like gold here, because collecting them comes at a high cost. Some of Norma's family have given their lives to find fuel. Lugging it back to the house is exhausting and fraught, but as essential as finding food. The stones used for starting fire—currently three pieces of quartz and a chunk of shiny obsidian—are sacred. You can't see them because they're too precious to leave in the open. Not that Norma's family can't be trusted, but they never know when they might get an uninvited guest. This is a relatively good neighbourhood, but if one of the other clusters of Homo sapiens (or indeed, other hominids) hunkering down in other houses lose their own fire-stones, it's mayhem for a while. Norma's house is guarded with stone axes, stones and heavy clubs. There's currently fourteen people living here, and there's never a time when everyone is asleep. It's not written anywhere (no written language for about 3,000 more years), but the rota of guards is never challenged. A system involving cooperation and reliability means the difference between life and death.

But I know where the firestones are. They're kept under a small hollowed-out rock behind that smelly pile of furs over there (*don't tell a soul*). Only Norma's mother knows the location because her job, aside from keeping Norma alive, is fire-maker. Every night of the year—even in summer—she lights the fire by which food is cooked. Mainly it's vegetables. Meat days are good days, and they come in clusters because the slain mammals are usually huge. If there's nothing to cook or it's not very cold, she still lights the fire. It's a ritual on which the day pivots. Fire offers warmth and a sense of safety—but not actual safety. Dangerous predators have become bolder, and fire no longer holds the same power. Who are the predators? Not the mammoth or the woolly rhinoceros or the elk or the straight-tusked elephants—although they are dangerous in their own way, they're herbivores and have no interest in killing people. No, the seriously scary predators are the scimitar-toothed cat, the cave bear, the cave lion, and the cave hyena. Norma will be terrified of these four animals when she's older, even though she'll have never seen them, because her family tells stories and draws pictures of them on the big stone which is the back wall. She'll have nightmares about a cave bear (Caba, she'll call it) coming through their door. She'll devise several contingency plans, like hiding, playing dead, or screaming so loud she scares it away. Her family has only one domesticated animal, which they share with their neighbours, a settlement of fifty-five people. The aurochs is a bigger, uglier, more aggressive version of our cow. It's almost impossible to tame an aurochs, and the only reason Norma's family have managed is because the beast's mother died when her baby was two weeks old, so it was easy to catch and tame. This is the way all domesticated animals begin—from matricide. Wild boars will be domesticated soon too, and pork will become quite a popular delicacy.

~~~

The mother unplugs Norma with a business-like gesture, puts her gently down on a skin-rug, then walks out of the house with a cranky look, like she's just remembered she forgot to take in the washing and it's starting to rain. I get a whiff of menstrual blood as she passes, so maybe that's why the cranky look. Not the washing.

Norma is content on her back and stares at me intently.

'You okay there, sweetie?' I say. 'Warm enough?'

'Waabababa,' she says softly, then blows some bubbles. I speak Bubble too, and reply in kind.

'Want me to hold you? That ground looks cold and hard,' I say in Bubble.

'Golagowawa,' she says, meaning if you touch me I'll bite your hand off. Or my mother will.

'Okay. No problem.' And then I ask: 'Are you happy?'

I've never asked an ancestor this question, and I've no idea why it pops out now. I've no particular desire to re-visit an ancestor who only speaks Baby, and maybe this frees me up. Like asking a bartender or taxi driver or anyone you'll never see again. It's exciting because, by the way she's gone quiet, I feel she's about to be coherent. That she's composing an intelligent response about personal happiness in the Neolithic period. I lean in and slow my breathing.

'Yes?' I coax gently. 'Take your time, Norma.'

Then there's the unmistakable sound of bowels moving, and a look of euphoria passes over her face. No words, not in Baby or Bubble. Just a full Neolithic nappy and a cloud of stink. Norma is just as helpless, and as easily pleased, as a modern one-year-old. In fact, more so. Her life span is shorter, and she'd be considered a late bloomer these days—not a good combination. I suspect it's brutal being one in 8000BC Britain. But now she's making a whiney mewing noise, which is a little grating. *Shut up,* I will her after ten minutes. Maybe I'm just tired. The grandkids stayed until late last night and I woke up with a babysitting hangover.

'Nanananana,' I say to her mockingly, stick out my tongue and then feel bad. I'm the adult here, after all.

'Sorry, Norma.'

But it's too late, and she starts howling. This has never happened to me before. It's distressing. I thought I was good with kids, but *she just doesn't like me*. What am I supposed to do?

'Listen, Norma. I'm sorry you can't talk to me properly, but I understand. I didn't mean to tease you. You don't need to like me. It's pathetic of me to want you to. I'm not everyone's cup of tea.'

'Mamamamam. Wuh!' Then some proper screaming. 'Wah! Wah! Wah!'

'Exactly! I agree, you're absolutely right,' I shout, but she won't be soothed.

Jesus Christ, I think. Not every ancestor is fun.

Florence

8000 years ago—5900 BC
What will be Doggerland in the North Sea

'Hey! Hi! Howdy do!'

'Not now, Florence. Not yet.' She's jumping up and down like a yo-yo. An early walker, is my guess, and chatty as hell. Refreshing after wussy Norma, the Neolithic nappy-filler. Florence isn't even wearing a nappy.

'Yeah! Gimme that!'

'No,' I say firmly, and put my hand behind my back. How does she know her gift is in it?

'Yes, yes! Now!'

'No! In a minute, alright?'

I like Florence, of course I do, but I'm going to ignore her for a few minutes because I need to backtrack to Eve again. Remember her? The aggressive female primate we met accidentally during a visit with Elizabeth and Ned the Hunky Neanderthal. Eve is the pivotal primate for all Homo sapiens. She doesn't get a whole chapter to herself because she's too scary and I'm a wimp, but it's impossible to talk about evolution and not mention her on a regular basis. Evolutionists use the name Eve for the obvious reason—they are mostly Western Christian scientists. (I happen to know her real name, or the contemporary equivalent of her name. It's Cruella.) The only way Eve's DNA will die out is if none of us reproduce or evolve into other life forms. Her genes, at least two of them, live inside each one of us—she is what we have in common. Which makes it simple, in a way, to picture our beginnings.

So forget bouncy Florence for now. I've finally learned how

to find Eve quickly.

Are you ready? Let's go! Hold tight.

There she is. Under that dead tree, tearing a leg off a boar-type thing that might or might not still be alive. Blood is spurting everywhere, especially it's running down her face.

Not pretty. But still, we owe her. We must applaud Eve. There she is, and therefore here we all are. Not a small thing.

'You again?' she snarls.

'Correct. Sorry. Sorry.'

'What *the fuck* are you staring at?'

'God! Not staring. Truly. Just paying my respects.'

'Huh. You hungry?'

'Sure,' I whisper. Did she just ask me to share her dinner? I'm not hungry, but I want to be polite.

'Well, tough. I ain't sharing. Now piss off.'

'Of course. I was just about to go. Sorry, sorry.' I start to back away, trying to remember why I came here. Oh yeah. Putting Florence in context. Then I pause and think. Post-Eve, but long before Florence, things started to resemble high school at recess, human-wise. Little cliques everywhere—some kids were jocks, some were brains, some were stoners. Some cliques were so tiny, no one knew about them. Some were so weird, no one understood them, not even themselves. No one was bonking anyone from another clique, or at least not very often and usually with a sense of embarrassment and desperation (like Elizabeth with Ned the Neanderthal). Little ghettos of related people were migrating all over the place, or—if genetically they had an aversion to the nomadic life—settling into fertile valleys to plant the equivalent of potato crops. They each became more and more distinct as they adapted to their place and diet, but all were Homo sapiens. They were mostly bonking siblings, cousins, parents, offspring. If you weren't prepared to bonk someone you were related to, you were out of luck. There are roughly 5,400 cliques and sub cliques today, and evolutionists call them mitochondrial haplogroups. It's a girls-only business. Boys get the gene from both parents but don't pass it on. Girls just get it from their mother and pass it on to sons and daughters. This is the second biological incidence I've noticed in which women have an edge, the first being that only women know who their kids are. Everyone in a haplogroup shares at least one ancestor, and there are not that many ancestors whose descendants hang in there.

'You still here? Fuck off.'

'Yikes. Sorry. Excellent idea, Eve, my fucking off. I was just thinking about haplogroups, which are basically subdivisions of all your descendants, and then I…'

'Are you *still* here? Are you deaf? I told you to…'

'Goodbye, Eve. Till the next time.'

'Not if I see you first, cunt face.'

Cunt face? Back to Florence, now!

Whoosh!

'Hey! Hey! Where did you go?' asks Florence.

'Nowhere. Anyway, I'm back now.'

'Gimme! Gimme!'

The gift is still behind my back and she's still hopping like a yo-yo.

'I said not yet, Florence. One more minute, okay?'

'You already said that.'

'No I didn't.' What I mean is, I may have said that, but I'm not really that kind of person. My kids might disagree.

'Liar, liar, pants on fire.'

Jesus Christ, I think. She reminds of Cora, her hominid grandmother times a million who'd never seen the world without snow. The similarity is not surprising. The genes that result in Enthusiasm and Unjustified Optimism are particularly resilient, as they both result in a robust immune system. When they don't lead the cheerful host into the mouth of a giant tiger (*What? You're not smiling?*) or worse. But back to haplogroups. Florence and I are, how shall I put it? Exclusive. We belong to the V haplogroup, which is very small. You're probably an H, along with 40% of all Europeans. The H's are the popular kids at school, a loud rowdy gang roaming the schoolyard and tossing their rubbish everywhere, whereas us V's are huddled in twos and threes and sometimes lonely ones around the periphery. We pretend to care about what's in our lunch box in hopes the H's won't notice us. Only 3% of people share the ancestor common to me and Florence.

And who is the original V? My great grandma times a million or so, Violet—that's who. Six thousand years before Florence was born, Violet lived on the Iberian peninsula about 1,600 miles south of here on the west coast of what will be Spain. She was amazing, was Violet. She had uncanny luck in staying alive, and more than that, she had excellent taste in men. She didn't care if they were ugly or old or couldn't tell jokes or, worse, insisted on telling jokes, but she wouldn't bonk a weakling or a simpleton. Her children, all thirty-two of them, survived to reproduce a dozen or so of their own. They had robust immune systems, fast reflexes, good eyesight and hearing, low blood pressure and a tendency to sleep well and eat anything. Even worms. (Sorry William, if that offends you.) The Iberian peninsula was a good place to live—further north, most of Europe was covered in snow and ice, deep in the same ice age we are slowly emerging from now. Within three generations, Violet had over a thousand healthy descendants, and it carried on from there. If a man from another group, maybe a wandering H, happened to bonk one of Violet's girls and she got pregnant, their child (boy or girl) was a V. It was a landslide of V-ism. I've visited Violet a few times, and I'm kind of a V groupie. She was something else.

~~~

A sound of throat-clearing. I look around and realise it's me. Florence is still at my feet, staring at me intently, and I've got a frog in my throat. I've got no idea what to say, so I say the first thing that pops into my head.

'Florence! Is that really your name? I had an aunt called Florence.'

'Yes indeedy deed. Cock a doodle doo!'

She crows like proper cockerel, then giggles hysterically. Not shy or serious, this one. Look at her. Red hair, blue-grey eyes, freckles. Spots of bright pink in each cheek. Aside from the mangy hair and her nakedness, there's not much to say she's not in my time. Actually, my kids had mangy hair sometimes and often preferred nudity, so the difference is minimal. Perhaps just a wildness in her eyes, beyond the more timid and self-conscious expressions of my offspring. I take a mental snapshot. If I had a pencil and paper, I'd try to sketch her. She's easily the cutest ancestor so far.

'Can I tickle you?' I say on impulse.

My fingers are full of tickles, but I'm mindful of boundaries. There has never been a time when personal boundaries didn't matter.

'No! Yes! No! Yes!' she says, extending her bare feet one after another, hopping around and giggling. I tickle them and she shrieks as if I'm killing her, then laughs like she's drunk. Or like a giddy toddler, which is what she is.

'Your hair is such a pretty colour, Florence. You remind me of one of my granddaughters.' I had red hair too, but I don't tell her that because it's not red anymore. Unless it's very sunny, and then it gets a kind of bronze sheen and people think I've dyed it (I have not!).

'I know! I'm the prettiest girl here.'

'My goodness!'

Pause while she considers me.

'You, however—not so pretty,' she says.

She tilts her head and smiles, as if she's simply making an observation. Not good or bad, just a neutral assessment.

'Yeah. Not so pretty.' I make a silly face to show I don't mind. Which is a lie.

'What are you?'

'I'm what comes from you, after 8,000 or so years.'

Short pause, while she digests this. Finally she gives a smile I can only interpret as pitying.

'Sorry, but I don't think so.'

'No?'

'No. I mean, look at me and look at you. Pretty and not so pretty.'

'Okay,' I concede. 'You're very pretty in Mesolithic terms.'

'Which means…what? Other Mesolithic girls are bit funny-looking?'

'It's all relative.'

She stares at me. My pun is unintentional, like all my puns, but still—I've made myself smile.

'What are you smiling about?' she asks.

'Relatives are relatively attractive. Get it?'

'No.'

I remember how much I dislike word play puns. They seem a kind of affliction or a form of showing off.

'Sorry, Florence. Ignore me. You are the prettiest girl on the planet.'

'I know! I just told you that!'

I crouch down next to her, look straight into her eyes and say:

'You're a V, did you know that? I could call you Miss Vee.'

It's unethical to impart knowledge from the future, but I can't resist.

'What's a V?'

'Good question,' I say, and unbidden into my mind comes Sesame Street. *Today's program is brought to you by the number 8 and the letter V.* Unlike all the other strands of DNA curled deep inside our genes, mitochondrial strands of DNA float free as a bird in the fluid beyond the gene nucleuses. And the ones that last from generation to generation get to have a letter. But this will not wash with Florence, so I condense it. She's playing with a piece of broken shell in the dirt. Making little roads and humming happily. Not looking at me, and no doubt borderline bored.

'Look, this is all you need to know,' I say. 'You and me belong to a teensy weensy family called V who all have the same great grandmother (times a lot), and her name was Violet. Most families are much bigger, but we're not small because we're weird. We're special. Like geniuses.'

I sound like my mother putting a good spin on my below-average report card. Am I channelling her? Of course! We all channel our ancestors because they made us out of themselves. It's inevitable.

Florence makes a scoffing noise and doesn't look up.

'If you don't believe me,' I tell her in what I hope is not too defensive a tone, 'you can travel forward in time and ask my mother Mrs Barbara Jones. *Cynthia is a genius*, is what she'll tell you, in complete denial of my D+ average. *Just look at the way she ties her shoes!*'

'It's not that I don't believe you *think* that,' says Florence in her most grown-up voice so far today. 'It's just that you're wrong.'

'Wrong?'

'Yes, wrong. There are thousands of us. I know about V's and we rule the world.'

'Ah. Oh good,' I say, having decided to humour her. 'I'm relieved. Thanks for telling me.'

'No problem. Want to get your own shell and play tracks

with me?'

~~~

So few of us. Why aren't the V's a more successful branch of evolution? After all, having met Florence and my mother, grandmother and siblings, it's obvious the V's possess many good things in terms of creativity, immune system, intelligence. Perhaps we lack sufficient competitiveness because we are so, huh, sort of *artistic*. So sensitive. So deep and honest (not). Or perhaps we just don't have a huge sex drive. Not because no one fancies us. No, no, no. That absolutely cannot be the case.

'Well! Anyway, Florence, I've come to say happy birthday. You've made it to one year, and the odds were you would not.'

'What do you mean?'

'You had a high fever and cough when you were a newborn, and there was a night when your mother assumed you would, well, die.'

'Yikes.'

'And then there was the time your brother dropped you in the river. Accidentally on purpose.'

'I don't remember that.'

'You were six months old, just starting to get cute. He was a little jealous.'

Pause.

'I like where you live,' I say, to change the subject.

'Me too,' she chirps.

Look around.

Rolling hills, pretty lakes and rivers, green fields and birch woods, smoke rising from teepees, and is that cow-type thing being milked by a woman singing a song about milk? It's peaceful. Life is easy here, like the Shire in Lord of the Rings before Sauron. No one here knows that Narnia is fast approaching. That they are living in the final days of the current interglacial period. But not everything is hunky dory. Some of Florence's family have recently moved to higher ground because the sea is encroaching due to melting icebergs up in Norway. Doggerland used to be the size of Germany, but in Florence's time it's really more of a wetland the size of Yorkshire. Not impressive, aside from this idyllic place, *the Shire*—which will one day be called Dogger Bank and become a popular place for offshore wind farms and fishing boats.

'At low tide,' explains Florence proudly, 'you can walk all the way from Norfolk to Amsterdam, right across Doggerland.' (Of course, she doesn't say Norfolk or Amsterdam or Doggerland, I've added those because the words she says would be incomprehensible to you.)

'Interesting,' I say. Though in my opinion, that doesn't make Doggerland a country.

It should really be called Dogger Island, I think.

Why hasn't Florence's family moved to higher ground too? Perhaps because the sea has been encroaching for thousands of years and it's easy to pretend it won't get worse quickly. Besides, look at their house down there by the beach. It's been in Florence's family for ten generations. They don't have a nomadic bone in their bodies and cling fast, even though every year they have to dry out domestic stuff and re-waterproof their house. This might explain the way in about 8000 years my Irish great grandparents will respond to annual floods in their mud house with such pragmatism. *It happens! Get over it!*

~~~

This is not the old stone age, no no no—this is the new stone age, a time of Neolithic societies and it's more civilised than I thought it'd be. They consider themselves incredibly modern, and like us, feel superior to previous generations. They weave beautiful cloth, sew treated skins into waterproof jackets, make fishing hooks and needles from bones. By the deer horn headdresses I can see hanging from a tree, it looks like they might have spiritual rituals too. They live on hazel nuts, berries, fish, red deer, and the occasional mammoth (weird elephant), which sees them through winter if they're careful not to get killed. Florence's house is round and six metres in diameter, with the same thick wooden supports her great grandparents put in place. The animal skin and rushes wrapping round the supports are watertight. The wooden struts of this very house are now in a museum near Star Carr in Yorkshire. They can be seen and touched. The remains of their boats can be seen too. Neolithic folk are not, *I repeat not*, primitive. Florence's family are very much our tribe. Maybe in the year 5000, Homo sapiens—should they avoid extinction—will study the twenty-first century and be amazed that we were not as primitive as previously thought. That we too valued spiritual rituals and domestic stability. Or maybe not.

~~~

But for right now, the world is ending again. I thought a tsunami would look like a big wave, but it's far stranger than that. The sea breathes in and in and in, and everyone waits for the exhale, for they've never known the tide to go out and not come right back in. Even the gulls act worried, circling and making baby-crying noises. Some people start walking out on the sands just because they can. Then, without any warning sound or sight, the whole sea rises about thirty feet—not in a curved wave shape, but all of a piece. A wall of sea water simply moves towards the land without fuss. No wild winds or crashing surf, just a rush of very briny air whooshing before it, as if the sea is relieved to let go after holding its breath so long. I'm suddenly above the water, treading air. In horror I watch Florence being swallowed by the sea. This seems to happen in slow motion, though it can't take more than a second. Her red hair is my last sight of her. All the houses are disappearing, along with 100,000 or so people. And their boats, their tools, their ancestor's graves, their religious objects, their weird cows, their flutes. Yes, their flutes. It's a nightmare even though I knew this was going to happen. I dog paddle in the air to the nearest land, which is covered with animals and people who've fled. There are bodies everywhere, but no one seems bothered. Maybe it's too soon. Maybe they're too frightened to be upset.

'Florence! Florence!' I call out, without hope. 'Florence!'

'Help! Help me!' comes a tiny voice after a while.

I breaststroke down to earth and land clumsily. I can't see her, but I know that was her voice.

'Help! Help! Help!'

I hear her voice as words but really it's more of a thin wail, almost a squeak. The kind animals make when they know there's no hope. The rabbit just before the car's tyres.

And there she is. Trembling, huddled, naked, alone. Snot running into her mouth. Wet skin coated in sand, her hair plastered down her back.

'Florence, I'm here. It's okay.'

She doesn't respond and I wonder if she's in shock. I kneel on the sand next to her and try to scoop her into my arms, but my limbs have no strength here. I'm doomed to stand by while she freezes to death, her lungs half full of sea water. Secretly, I've always wanted to rescue a foundling—but now I've finally got one, I've no substance with which to carry out the rescue. Yes I can carry cakes and lend cardigans to fruitarian tree dwellers, but I cannot lift a child. I settle down as close to her as I can. Fifteen minutes go by, it seems like days. Her eyes close and her breathing slows. Her skin is blueish but she's still breathing. I remind myself that death can be like falling into a dream. She'll not know, but this is not much comfort. For no reason—aside from the quirk of thought processes in general, the way they work at their own pace while the rest of the mind is distracted by more immediate concerns—it suddenly occurs to me that the Doggerland drowned are mostly V's like Florence and me. Of course they are! She was right, I was wrong. The highest concentration of V's in my lifetime is along the coast of northern Germany, as well as eastern Scotland and parts of the English coast. The descendants of the Doggerland tsunami survivors. V's are not few in number because no one wanted to bonk them. They just drowned. This makes my ego feel better, but only for a second.

I feel certain Florence is dying, though I am equally certain she is not—for otherwise how can I exist? This sensation of being drenched in contradiction is extremely familiar, so doesn't bother me over much. It's normal to feel torn. *She cannot die*, but even so I don't take my eyes off her in case the act of staring is keeping her alive. Who knows the truth about such things? People are walking around, mostly alone, shivering, dazed. Then an old woman approaches. Well, she might be old—hard to tell, in the dimming light. She's wearing some kind of fur, wrapped around and tied at the waist with something that looks like dried intestines. Holes, out of which poke her skinny white arms. Her hair is dry, as is her fur coat, which seems extraordinary. Maybe she found a tree to sit in? She approaches Florence and stops. She's curled up in a ball on her side now, with her face away from us. The old woman nudges her back gently with her bare foot. It's a repulsive thing, long toe-nailed and black with mud. Florence does not respond. That lively little ginger nut is dead, dead, dead. The old woman nudges again, but this time a little harder.

'Stop that,' I tell her. 'Leave her alone.'

But of course she can't hear me and just keeps nudging.

Florence's body flops a bit each time she does this. I can hardly bear it.

The old woman turns to continue on her way, peering here, there and everywhere. What's she looking for? Is she a Neolithic looter? I'm relieved she's gone. I trust no one. Then she turns around abruptly, returns and squats down next to Florence. *My* Florence. Now we're so close—hardly an inch between us—I notice how disgusting the old woman smells. *Cooties!* I think in an infantile reflex, and automatically breathe more shallowly. She takes off her fur coat, lays down in the mud next to Florence and tucks the coat around them both.

Oh my.

Something so timeless and poignant about this pose. I know this sounds patronising, but I didn't think kindness on this scale had evolved yet. I thought it was a luxury trait for more modern times. The old woman lies like that for about ten minutes, rocking slightly and making a steady guttural noise. No words, or not in a language I understand. Which, given my comprehension abilities so far, seems unlikely. Then a cough. Florence coughs! A tiny sputter, and then she starts crying hoarsely. The old woman shifts her around a bit, and after a while I hear the sound of hungry sucking. Okay. Not so old. A mother, with perhaps a drowned baby.

~~~

And that, I think to myself, is the happy ending to the story of Florence. If she'd not been found by someone with active milk ducts, I'd not be here. Thank you lady, whoever you are. And as in a fairy tale, everything repulsive about her melts away. She smells sweet, her black toenails are delightful. My time here is done and I feel myself thin out. I'm porous, light as a feather. I take one last look around Veeland, as I now think of Doggerland, and even though all I can see is the reflection of the new moon on the sea, I make a solemn bow. Britain is now an island, and most of my peeps are at the bottom of the sea. My beautiful drowned V's.

Tony

1067 BC
What will be Italy

The Stone Age bit the dust about two thousand years ago, followed by the Bronze Age which wasn't too bad if you like bronze. Now we're in the final days of the Iron Age—*check it out*. Iron has been making all the difference because stone, copper and bronze make pretty useless tools, always bending and corroding. Iron farming tools and weapons—now you're talking.

According to some twenty-first-century scientists, we left the iron age about 550 BC—although others speculate it lasted till AD 800. Eras are nebulous things and tend to be named as such long after they have passed. Our current era is officially called the Modern Age or the Modern Period—which I suspect everyone has called their own time since the beginning. Meanwhile, regardless of what we call our age, we're living in the final years of Earth's fifth ice age. I have to keep reminding myself of this. Everything is so different from Florence's cataclysmic time, but still—everything is kind of the same. The world keeps almost ending, and then it keeps lurching forward. It'll be three thousand years before my conception. All those tenuous connections yet to be made in order for me to be me, and you to be you. If I go back only forty generations, I have roughly two trillion direct ancestors. If I go back another forty, I have four trillion ancestors. Between me and Tony's time, I have more grandparents than the entire current population on the planet, which is 802,014. As I write this book, there are over 8,000,000,000 of us.

This boy in front of me is Tony and he's one year old today. He gets a place-appropriate name, unlike previous ancestors—although in truth, it is still the equivalent of his real name, which I have no access to. There is no Italian language yet. He's my g g g g g g g g g g g grandfather plus some. Yup, I'm roughly .5% composed of this twenty-four-inch tall person.

Look at him.

Isn't he adorable? Sturdy and curly-haired, like a miniature man already with a little pot belly. Like Bam-Bam from the Flintstones. Subjectively the most aesthetically pleasing male ancestor yet. Despite their future reputation as ignorant and superstitious, no one here thinks the Earth is flat. There's no heaven or hell yet, no devil, no saints. (Talk about superstitious!) Jesus Christ—who even from an atheist point of view was at the very least an extraordinarily decent human being—won't be born for another millennium. We're in a place that won't be called Italy for five hundred years—by the Greeks, ironically. There are substantial towns with well-constructed buildings, and not a single person here considers themselves old fashioned. In this house—Tony's home—everyone believes that everything around them is inhabited by spirits or gods. Everything and everyone. The cup they drink from, the water in the cup, their throats which swallow the water. Gods of unknown origin who have mysterious powers and are not always fun to hang out with. They also believe the spirits of their ancestors are everywhere. Dead people are never completely gone. Ancestors are watching, influencing their lives. The air in Tony's home is full of the unseen. Is that so hard to imagine? Our air is full of emails, texts, photographs, streamed movies, voices from thousands of miles away. A multitude of gossamer strands with substance—otherwise why would a wall prevent decent internet reception?—and yet while travelling to us, they are as invisible, intangible and inaudible as Tony's spirits. Everyone here in Tony's town believes spirits exist with the same confidence we *don't* believe they exist. Which begs the obvious question: Which of our current pieces of certain knowledge will one day be considered ignorant and superstitious? No one knows these things, but understanding this, I approach all certainties with caution.

Oh, just look at this boy!

So clever, the way he hardly dribbles when water is offered to his lips. Such serious dark eyes under such lush dark eyebrows. I ache to touch him. It's hard to tell who his mother is. He toddles to a young woman who is holding her arms open, then he swivels, falls neatly on his bottom the way one-year-olds always fall. Gets up and toddles to another woman who is also holding her arms open. I can't see any men, but I hear men's voices outside. I don't understand the language— it's not Italian or Latin—but I can hear some animals—cows, I think—sounding distressed. Maybe they're being herded by the men down the road to market while the women stay inside and do house things. There are other children, five or six. One is younger than the birthday boy, and the others all seem to be about five. Chubby cheeked, barefoot, scabby kneed. Someone—a servant girl? An older sibling?—is passing a tray of food around. Dollops of brown and black, which everyone seems to love eating with their fingers. It smells a bit gamey. Maybe stewed rabbit.

'Happy birthday, Tony,' I tell him.

I give him my warmest smile, which is not an effort because I already love him. It's incredibly easy and feels great. (Loving is lots better than being loved. I have no scientific evidence for this, but that doesn't mean it's not an empirical fact. Sad and feeling flat? *You better find somebody to love*, in the words of Jefferson Airplane.) No answer. He doesn't look at me. In fact, he looks a bit gormless, and because he is related to me, I immediately edit my belief that high intelligence is superior to low. How can anyone be fairly judged by a quality they are born with? I've been a snob. I blame Freddie in Kenya, the red oblong fruit-eating hominid. That's when I began to assume being smart was something to brag about.

'What a big boy you are!' I inject some enthusiasm, in case I've underappreciated him.

Still no response. I stand directly in front of him and make a funny face. Not even a flicker. This is new. Am I invisible here? Not good news. Who am I going to talk to? I feel lonely already. A bit silly and superfluous. Then I fart because I had beans for lunch. Also because ancestor visits give me gas sometimes, I don't know why. Tony laughs in a way that only a fart could elicit. The fart laugh. Aha! So, he's the anti-Florence. Shy, inarticulate. That's okay. Conversation is overrated, in some ways. Much less exhausting, to not connect verbally. I find a place on the floor to sit, surrounded by the hive of domesticity. I sit there for a few days, which pass the way my days tend to in the distant past—painlessly, with no hunger or thirst or need to use a toilet. (At home my house is empty, my husband having gone to Majorca with his daughter for a week. Hence in theory I'm free to stay here till Friday afternoon.)

Overall, it seems to work, this household of women. There's always freshly ground flour in the flour barrel. There's always fresh cow milk and sometimes ewe milk too. The children don't seem to form particular attachments to women based on relationship. All the women mother all the children, and babies gladly latch on to any proffered nipple. Actually, even the five-year-olds latch on. After a while, I find the room soporific—the constant soft chattering and tidying up, the crying and consoling, the chewing and swallowing and sucking, even the sweet smell of baby poo and breast milk. Nostalgia washes over me. I remember watching my own babies toddle towards my open arms, of breastfeeding them, of washing poo off their beautiful bottoms. One-year-olds poop wherever they happen to be, no matter what period in history it occurs, and breastfed babies only make pale yellow stink-free poops. I'm content here in this Iron Age house. Even so, this is a lonely place for me. Tony is my direct ancestor but he's not talking to me. Pointedly not looking at me either, in the way shy people keep their eyes averted. And I've run out of farts.

Oh well.

Eventually I stand up, have a good stretch and stand by a glassless window. Because I'm feeling invisible, I burp loudly and scratch my armpit. I pick my nose and study my findings. I comfort myself by considering the bigger picture. (It's hard to be cocky, or sad, when you remember the bigger picture.) It goes both ways, evolution—my scant beginnings and my minute unravelling. The sources of me and the results of me, on and on as far as the eye can see in both directions. (I assume. I have no access to the time beyond my own existence.) If you do not reproduce, you are still part of this chain, for whether you choose to or not your existence affects others who in turn affect others. Having sex that results in an egg being fertilised does not require intelligence or courage or integrity or even intention. And a child growing into a thriving adult does not require a biological parent. We are all capable of ensuring the longevity and well-being of our species just by being good people. Sounds simplistic, but the kind of daily self-scrutiny needed for good behaviour—especially for kindnesses which are rarely noticed or rewarded—this effort is not simple or easy. I've been visiting ancestors for a few years now, and I can tell you this: No one manages to be good all the time. Not a single life form. Not many even consciously try, although most are instinctively good. Which is a relief.

In a DNA rewind sense, we are all here from the start (*thanks be to Kevin, amen*). The first pulse, the first inhalation, and we've believed in countless entities and called them a thousand things. In the last two millennia, us westerners condensed the gods into God, a male of indeterminate age, appearance and address. We've prayed to him, given him (Him) a capital letter, fought for him, but increasingly we disbelieve in him. If he's here today, it's not likely our rejections bother him, being God (plural or single, male or female). If he's not here, there's still the mystery of creation to ponder and even worship. What set the whole show going? And what set that show going? When I'm in this kind of mood, the past feels like rows of teeny dominoes, all falling in a direction which miraculously ends up with me being me and you being you. If, like Carl Jung, you think you were meant to be and all coincidences have meaning, you might call it synchronicity or fate. I do not. I see myself as a fluke. I could very easily not be.

Which brings me back to the miracle of brooding Tony.

~~~

He was born in the room next to this one. The town is nestled in a valley with hills surrounding it like sentinels, and half a mile away is where my great great grandmother's mystery father will be born in almost 3000 years. For a short time, he'll live in this very place—not the same building, of course. This house will be rubble in the blink of the eye. He'll go north for work and stay long enough to impregnate teenager Maria, before heading south again because he misses the sun. He'll never know he's left a daughter behind who will grow up in an orfanatorio run by Sister Rotund, and that this daughter is my great great grandmother Giulia Cardone. His name is *Randazzo Zacco*. For the sake of this wonderful name, I forgive him all. My razzamatazz grandfather times three.

~~~

No one in my family will discover Tony's name or birth date. It will be over three centuries before any of that kind of hoo-ha begins, registering births and deaths, and another thousand before it becomes common practice. Unless, of course, you are a king or queen or a marauding invader alpha male like Attila the Hun or Genghis Khan. For now, there is only time for staying alive and no purpose in recording dates. Although Homo sapiens have been writing for a few thousand years in places like Mesopotamia, not many people write here yet. A few are keeping diaries, but none of these will survive. In Tony's time, about a third of all infants die at birth or soon after. Lots of people die before they hit twenty, but the ones who make it often live to the ripe old age of forty. Days are for lurching through, a constant effort to minimise chances of premature death. But I've noted that the inhabitants of this house don't seem worried. Probably hard to sustain intimations of mortality when the sun is shining and your belly is full. And after all, daily danger is easy to ignore—it's mundane. You still have to do all the ordinary routine things, no matter what. Eat, drink, pee, poop, get dressed, get undressed, wash occasionally, sleep.

~~~

Now I'm freed from interacting with my ancestor, it's easier to look at his future without worrying my absence might bother him. Various events—violent and political—are already leading to the rise of the Roman Empire, but no one around here notices. No one here has been to Rome or even heard the word because it doesn't exist yet. Theoretically, little Romulus will come along in about 500 years and kill his brother Remus. Both having been raised by wolves, this is probably inevitable, but still. *Kids!* Romulus will establish the city state of Roma on Palatine Hill and commence colonising the surrounding world with astounding success. But Italy itself will not be known as such for another 2500 years. Tony's descendants (Yes, of course, shyness does not preclude virility. Not even gormless-ness does that.) will be citizens of the richest state in the Roman Empire. Until, like all empires (like China and Greece and Russia, like Britain and other northern European countries, like America) Rome will eventually fall. Yes, the crazy Visigoths from Germany hasten the end, but essentially it implodes from within like all empires. Oh, the fatality of affluent arrogance! Of complacency and greed! Sadly, this is a lesson Homo sapiens seem incapable of learning. Or they learn it, but their territorial instincts override it. Luckily, so far, their self-preservation instincts keep overriding their territorial ones in the nick of time. But it's a risky dance, and getting riskier.

~~~

Marriage has been around for more than a thousand years. It helps to keep society glued together and babies tend to survive longer. In fourteen years, Tony marries Luciana. (Luciana is one of the babies sitting in the first birthday room, and who also appears to need her bum cleaned.) The ceremony is short and afterwards they have intercourse in the room he was born in. People listen and laugh at the sounds from the other side of the wall. His mother prays loudly to Bona Dea, goddess of fertility, for her daughter-in-law to get pregnant, and also a silent prayer to Liba the god of virility for her son to perform well. He's been such a worry, that one, with his silence and slowness. Luciana falls pregnant immediately, then loses the baby before it's born. Their second baby dies at a week old, is wrapped in a shawl and buried in a quiet place where the gods can look after him—a place reserved for newborns and children under one. The innocent ones. Only the grandmothers cry. The third pregnancy is easy and the child—a boy—is strong. His name is Luca, and he's my grandfather times a hundred or so.

~~~

Tony works hard. Unlike his wife and everyone else in this house, in fact almost everyone else in Italy, he hardly talks. Sometimes days go by and no words leave his mouth, just quiet grunts. His eyes are quite eloquent, as are his beautiful broad hands. His job, along with his father and brothers, is to use an iron pick-axe to hack marble and other stone out of the hills, then keep hacking smaller pieces until the desired shape and sharpness is achieved for tools and weapons. Or the right squareness for building houses. Or flatness for paving roads and paths. The work is never-ending and he's been doing it since he was seven. His mind empties as he works, and his days are punctuated by meals and occasional sex. He is fond of his wife. She is sometimes cranky with him, but there is no brutality or animosity between them. (There's a lot of that around. Maybe semi-muteness precludes aggression? Maybe the energy required for anger is too much?) Nor is there any infidelity between them. She often prepares a meal for him to take to work and wraps it in a waxed cloth, which he folds into his pocket and returns to her in the evening. He has a particular fondness for pickled eggs. He always eats with a steady concentration, if not pleasure, before closing his eyes and falling into a fugue-like sleep in the places he always sleeps. A normal day for Tony contains three post prandial naps of about ten minutes plus ten hours at night. He always dreams, often of the wild mountain horses he saw once as a boy. I don't have access to his dreams, but once he mentions them to his wife.

'How did you sleep?'

He nods once, half smiles, says 'Horses,' and then gestures to the mountains.

'How lovely, lucky you,' she says. 'I only dream of a floor that does not need mopping.'

As far as contentment is possible, Tony seems to possess

it. How do I know? Because it's easy for me to slip into the consciousness of my more recent ancestors. I don't even have to try. It's an effortless empathy which is not always rewarding, of course. I mean who really wants to feel how someone else is feeling? It can be exhausting and even scary. Though the inner life of Tony is relatively pleasant to dip into so far. Currently I feel contented.

~~~

Time passes. He and Luciana have ten more children, seven of whom survive—above the odds. His eldest, Luca, joins his dad and uncles at the quarry when he's seven. Tony's brothers' wives, who also live here, have more babies. The house is now four times the original size. Extensions grow on extensions. There are now four storeys with steep stairs and sometimes ladders connecting them. One early morning of a full moon day, at the grand old age of thirty-eight, Tony tells his brothers:

'Not going to the quarry today.'

It's a long sentence for him, and no one objects. Tony is no shirker, so he must have a valid reason. He indicates he isn't feeling well, touching his stomach and frowning—but this is a lie. The truth is, he just woke up in a mood. Maybe wanting some time home alone. It's so rarely quiet in this house. The women (all seven of them) and children are getting ready to leave for the market, baskets and babies and pots strapped to their backs. They have no money in the modern sense, nevertheless they will return with the needed objects inside those baskets and pots and feel part of the modern world. Cheese in return for eggs, cloth in return for meat. One of the women has an ochre-dyed woollen shirt that took her a month to weave, and she's hoping to return with a kitchen pot made of iron.

Luciana says he should come with them, and the other women chime in. He's always been popular, for unlike their husbands he never shouts or says mean things. They envy Luciana and consider Tony their teddy bear.

'Come on! You can taste Fiorella's new honey!'

'Be a woman at market for a day, we won't tell,' entices his wife.

He smiles—their excitement is contagious—but shakes his head.

When they're gone, after a short while the house spirits and gods emerge. He is not alone after all, but this kind of company is refreshing. It asks nothing of him but his awareness. He slowly swivels in the room, nodding a little, acknowledging the spirit in the wood of the chair, in the leather of his sandals, in the stone of the walls, in the lime wash on the stone, in the tomatoes lying on the window ledge ripening. The spirits of his parents and their parents and all their parents. They swarm around him in a comforting way, inviting him to consider them further. To converse, if he so wishes. He does not because he is cloudy-brained Tony, and that is alright. The spirits keep him company anyway.

He stretches, sits on the wide window ledge, leans against the wall. Then for a moment or two, he's happy and the world is amazing. But fast on the heels of his happiness comes a howling loneliness. He's always been a contented loner, so loneliness is a novelty and he doesn't know what it is. Maybe he really is sick? He gazes out the window with unfocused eyes, not really looking at anything now. He has a long dark beard and his long hair—not grey, but thinning—is pulled back into a ponytail tied with a leather strip. The muscles on his bare arms and legs are solid, even his substantial belly is firm. There's no glass in the window. Although glass has been around for a thousand years, it's not much needed here. The shutters are flat against the inside wall, and the sky is already a hard blue. He continues to sit, experiencing this new sensation, and listens to the distant market sounds. The call of barkers (*Half price goat livers! Going quick!*), the moos of cattle, the clatter of wheels over cobblestones, the occasional squeal of a pig, some pottery breaking followed by a woman cursing—and suddenly he can hardly bear it because he's woken up to the fact of his own existence in the world. Most people get to this place in early childhood. A few people never get to it, sleepwalking from birth to death. For Tony, the moment of awakening is now. The world is beautiful but he cannot bear to feel the beauty alone. His heart beats faster and he feels capable of great change. Of great clarity. He doesn't need his days to be the way they are. He doesn't need to stay shut up inside himself.

He reaches for the house spirits and gods like a person might reach for a mobile to ring for help. An automatic gesture—come quick!—but they are....why, they are not here! He frowns, summoning them with all his senses, but nothing answers him. Maybe they've ceased existing. Maybe they never existed and people have always been alone and that's all there is to it. It's August 14th, 1029BC, and a warm day—but now Tony is shivering. Feels hollow and frightened.

Oh! Oh! Oh! he whispers. And then a few *Huhs!* He tries to imagine talking to Luciana or his brothers later about the missing gods and spirits, about the way he feels—but he can't.

He hears the distant whoop of a man's careless laughter, and thinks without words:

All I ever wanted was to be like everyone else! To laugh like that.

Heaviness creeps into all his limbs, into his eye lids and knees and toes. Into his lips and ear lobes. Biological imperative of loneliness? Theoretically to ensure we work together cooperatively for a stronger society and increase chances of personal and species survival. Though Tony was doing that quite happily before today. Maybe loneliness is so painful it impels us to overcome it—to learn true autonomy, which is an excellent arsenal in anyone's survival kit. I'm only a bit ashamed to admit I've flunked at autonomy and lowered my standards multiple times to, as Kris Kristofferson says, help me make it through the night. (It was a long time ago, kids. You can stop cringing.) It's possible I married to avoid being alone. This does not preclude the presence of love, of course. Just makes me less noble.

But maybe loneliness is not the cause. Maybe Tony's dark mood is the result of the full moon. A sudden onset depression. Gravity pulling on the fluid in Tony's body. The cyclical fluctuation that we all experience twice daily without being aware of, and which increases during the full moon. More than half of Tony is water. How can he not respond to the tug of the moon like the tides? But why oh why is today different from other full moons? He's lived through 456 of them so far. Perhaps it's a combination of factors. The novelty of loneliness, gravitational lunar pulls, what he ate last night, minute hormonal and metabolic shifts, the fact he had diarrhoea a few days ago and has not fully rehydrated yet, the slight headache behind his eyes.

Shush. Now he's thinking without words: *Why didn't I go with everybody today?*

With every atom, he regrets his decision to not go to work or at least the market. Biological imperative of regret? And it's close cousin, guilt? After experiencing regret, you tend not to repeat the events leading to it, which is good. Unless you forget to not repeat that thing, and end up regretting all over again. I regret so many things, it's embarrassing. At three a.m., they swell and multiply and taunt. *Thought you knew what you were doing? That you were being smart?* Then I get up in the morning and try harder. Likewise, in better future circumstances, Tony would be unlikely to skip work or turn down offers of a trip to market with the women and children.

Damn, damn, damn! he thinks now, angry at himself.

He considers grabbing his shoes and running to the market, of joining his wife, but is rooted to the spot. Paralysed by his own nature. He can see another way of being, even see how much more it would suit him, but it's like viewing another person entirely and he cannot go there. He simply cannot. Suddenly he fills with loathing for the sound of his own breathing, the beating of his heart, the feel of his own skin. It's unbearable to be alive and not fully living. He swivels so his legs are hanging out of the window. Looks down to the street four storeys down. Has a vision in which he swoops over the rooftops and cobbled lanes to join the women and children and taste the honey and be part of the world. Then, without knowing he is going to do it, he lets himself go. It takes a twentieth of a second for him to change his mind, and he claws the air with legs and arms in an effort to retrace his way back to the windowsill. After all, most things can be reversed, right? And this is such a minute fraction of time, and such a forgivable impulse. He's travelling at seventy-nine miles an hour and accelerating by the millisecond. It doesn't take many more fractions of seconds to hit the ground—just enough time for his ears to register the rushing air. No last thoughts or images, not even a flicker of those wild horses loping.

Oh, Tony. Tony, Tony, Tony.

Self-destructive urges can be due to long term illnesses like depression or schizophrenia. But they can also be spontaneous and inexplicable, even to the person afflicted. Like healthy cells suddenly misinterpreting a bio-chemical signal and attacking other healthy cells—*mutiny on the Bounty!*—resulting in an autoimmune disaster, sudden suicidal inclinations can sabotage an otherwise healthy life.

Faulty assumption number one on which we rely anyway: *No one wants to die.* This isn't true, but we live as if it is because to do otherwise would mean major changes in our lives. Suicide bombings and suchlike are doubly disturbing. There's nothing we can do to prevent this kind of tragedy, because the basic self-preservation premise has become null in that person. People who do not feel the need to stay alive can make us nervous, because they have nothing to lose and therefore none of the usual social rules apply to them. We imagine they are capable of driving their car right into ours. That they are full of rage. Or a deep irrational depression, which might be cured if only they'd tell someone. But sometimes they just experience a sudden excruciating form of loneliness or regret or emptiness or self-loathing, and they have no defences for it. There's no biological imperative to this kind of impulse. The species does not need it to survive. Most likely, it's a random glitch—like the one toaster out of 271,788 that shoots out flames when you just want toast. Sometimes suicide is just as accidental as turning the wrong way up a one-way street or catching a nasty flu bug from some sneezes on the bus. A momentary blip, which if the person is lucky, passes without anyone being hurt.

When Tony is discovered a few minutes later by the goat boy who's always late to the market, at first he think someone has dropped a heap of old clothes in front of their building. Then he sees it's a man and he recognises Tony. Everyone around here knows Tony. He'd be amazed at his popularity. His limbs are twisted, but the side of his face that can be seen looks peaceful. Like he's taking one of his naps. I'd cry now, if I was a crier. Instead, here I am. Writing him a goodbye letter.

I'm sorry it ended this way.

~~~

Back to a happier day. A day which, to baby Tony, seems endless and full of newness. From dawn to dark, so many discoveries and pleasures and frustrations. Look at him, wriggling away from the woman holding him in order to crawl across the floor towards some scraps of food that have fallen from someone's dish. Nothing in his mind but anticipation of something savoury on his tongue and the sense of security that comes from being in a familiar safe place and having a full stomach.

'Happiest of birthdays, Tony,' I whisper.

I can't help trying one more time. I get a head tilt, but still no verbal response or eye contact. He doesn't know he's one today. He doesn't know he's going to marry Luciana, or be a father, or bury babies, or work so hard his hands bleed, or be hungry sometimes, or dream of wild horses. He doesn't know one August day he'll crave solitude and experience loneliness and therefore regret for the first time. That this regret will squeeze his heart and throat until he can hardly breathe. That even though his demise will be self-imposed and violent, he'll fall into it as gently as a feather into water. He knows none of this. Essentially, there's nothing about Tony today that isn't true of every one-year-old. All he cares about is getting that scrap of fried pig skin into his mouth.

Recent Homo Sapiens

Lady Margaret Drummond

1476
Perthshire, Scotland

Jesus Christ, so much has happened since Tony. Jesus Christ himself, for one thing. The Renaissance. The Roman empire expanding, contracting and finally imploding. The Greek civilisation rising and falling. The Great Famine and the Black Death, both of which are ongoing, culling Europeans at a rate of one out of two. The Dark Ages are coming to an end, which sounds like a good thing…but it also heralds the beginning of the slave trade, though of course enslaving weaker humans is old hat. It's just more organised now, more profitable. In fact, it's politically correct in the context of its times. Beware of over-judging in case it blinds you to your own misconceptions. It is a certainty that a future generation will accuse us of immorality and wickedness and stupidity. *Oh yes they will.*

~~~

It's unfortunate that Margaret has the same name as my ancestor who crawled out of the sea—in fact, I have 286 direct ancestors called Margaret—but *c'est la vie*. This is not a novel, and therefore will seem poorly contrived and confusing at times because real life is both those things. At least this Margaret gets her real name, not a generic equivalent. And she gets a surname, unlike the previous Margarets. No, I have no idea why. Maybe because only life forms with documents that have survived, earn them?

I've visited this Margaret before, of course. I keep coming back to see her because I feel so sorry for her. I mean, what did she do to deserve her fate? Also because I'm a groupie and she's my only relative who made it into Wikipedia—if you don't count her boyfriend. (I'm related to Roy Rogers too, but he's just a cousin—so no visit to his first birthday. Wah!) But my Margaret Drummond is a direct ancestor and she's famous. Not for doing anything remarkable and the price was obviously too high, but still. Though I'm sure she feels, if she feels anything now after five centuries, that fame is overrated. *To hell with thou, Wikipedia!*

Today is her first birthday. We're in Drummond Castle, near Crieff in Perthshire, Scotland. It's a dark and draughty place, with narrow staircases and smoking fireplaces and mouldy walls. In the dim damp kitchen, the cook is considering her available cake ingredients, for Margaret is my first ancestor to merit a proper birthday cake. Eggs, flour, butter, sugar and raisins. A pinch of pricey cinnamon if the cook can lay her hands on some. Margaret and her two sisters are lucky. They get to live in a castle with servants because their father is John Drummond, first Lord to King James III. John dotes on his girls—all of whom are pretty. Prettiness counts for a lot, for daughters are chiefly valuable in terms of beneficial marriages. Lord Drummond has high hopes for one of his girls to marry the next king of Scotland, who is currently three years old. He has to find upper class husbands for them all. Hey ho, I think.

These are the days of unapologetic class systems, unlike our more politically correct times of denial and middle-class guilt and double standards. In the fifteenth century, there are the monied and landed, and there is everyone else. People breed with their own socio-economic kind, and generally accept their class inferiority or superiority without much ado. No one expects Life to be fair. There is always someone below and someone above, and that's that.

~~~

Let's go up three flights of stone stairs, very narrow and very windy. Do not look down. Here we are. This bedroom is freezing and the floors are hard. There are tapestries at the windows and on the walls, and the air is heavy with melted beeswax and human humidity. Margaret has a mother, but the person she's mothered by is Liza Todd, her nurse. Liza loves her charge. You can tell by the way her eyes are following Margaret right now.

'Where do you think you're going, Missy Meggy?'

Margaret giggles—a naughty toddler chortle, and carries on towards the door.

'You little monkey!' scolds Liza Todd, but not as if she's angry.

Margaret has straight black hair, what there is of it. Her navy blue eyes are so dark they look black too, but her skin is extremely pale and already prone to blushing. (Biological imperative of blushing? It signals weakness, elicits sympathy, thereby reducing chances of being eaten.) It's a startling contrast, her black hair and white skin. At the sound of Liza's voice, she pauses in her rapid crawl to the door leading to the spiral stone staircase—where she is forbidden to go. She knows this, but (or and) the pull is irresistible. This will be the death of her one day. Not the staircase but the inclination to ignore Liza's advice. To be drawn to the forbidden. Biological imperative of the risk-taking gene, closely linked to the courage gene? Increased chances of survival, which are almost cancelled out by increased chances of early demise. Risk taking and attraction to danger are qualities associated with the world's most successful individuals and the majority of prison inmates. It is not known whether these inclinations are genetically on the wane or are waxing. It is known that in these relatively peaceful and safe times in which manhood rites of passage are absent, adrenalin sports have never been more popular. Nor have sports like football had more passionate fan bases. Male Homo sapiens apparently need to experience physical danger vicariously or in person on a regular basis. Some women (like Margaret), as well. The proportion of female risk-takers may be on the increase—but it's a gene that's still predominantly obvious in men.

~~~

Of course evolution is still quietly happening. And social evolution has begun to speed up, which also impacts on our physical evolution as a species. Since Tony's time, resistance to disease (passed on in DNA) has become more efficient. Body size has marginally grown due to improved diet, but weirdly, brains are shrinking. Not much, but some. Maybe due to a lessening need to store information as our social groups become more active in specialising and holding complex information for us. We have doctors for questions about health, police for questions about protection, priests and philosophers and evolutionists when we want to know what it's all about. (Where will it end, with our current dependency on Google and Wikipedia and AI's in addition to our professionals? Will we end up pinheads? Will large-headed humans be considered freaks and un-marriageable?) People are still moving around the world in the fifteenth century, but mostly they are not. They remain in the place they were born and make babies with near neighbours. This results in similar-appearing populations, which in turn gives birth to racial stereotyping. Therefore fifteenth-century rich families in cold climates tend to produce pale tallish healthy people with a decreasing likelihood of contracting diseases like typhoid or leprosy or malaria. People like my Margaret, for instance.

~~~

'Happy birthday, Margaret,' I say, without hope of being heard. Since Tony, I've accepted that ancestor-chat might be over. But she looks at me instantly and gives me an enormous cheeky smile. Maybe Tony had something wrong with him I didn't spot? Or there's something wrong with me, and he just didn't like looking at me?

'You can see me?' I'm suddenly aware I'm still in my bathrobe, having assumed I'd be invisible. Does it matter? Not a bit, I tell myself. It's an old man's robe—not belonging to an old man, but both old and bought in the men's department of M & S. Huge, saggy and grey. Not a flattering look.

'Course I can see you,' she chirps. 'You're standing right in front of me, ya daftie.'

She has a slight Glasgow accent, which is entirely inappropriate given her location and era, but makes more sense than some of the previous ancestor accents. Consider hominid Freddie talking like Stewie from Family Guy.

'Huh,' I say.

'More to the point, why are still in your robe? It's lunchtime.'

'Huh,' I say again, wanting a strong coffee. Or a gin and tonic.

Her darting eyes look me up and down. She frowns and says: 'What are those brown marks on your robe?'

'Uh, coffee stains,' I mumble, blushing. Then I gather my wits. 'This is not my robe, young lady. This is my, my, my work uniform.'

'Liar.'

'Not.'

'Are.'

'Okay. You're a smart cookie. You're right, and I'm just a lazy slob.'

'Cool.'

'Are you a lazy slob too?'

'Hell no. I'm pure dead brilliant, so I am. A real go-getter. You won't catch me in my robe midday.'

'Well, you better not tumble down those stairs or you'll never get big enough to wear a robe.'

She immediately revives her crawl to the door, but Liza Todd is there before her and scoops her up, laughing.

'You're such a little madame! What am I going to do with you?'

~~~

One May day, in about nineteen years, not far from here, while walking with her sisters through their favourite bluebell woods (their white bosoms spilling over their stiff bodices, their tiny feet in impractical satin slippers, their white hands fluttering embroidered handkerchiefs), Margaret meets the king of Scotland, as one does. James IV is twenty-two, two years older than her. He's been king since he led the rebellion that killed his father at the Battle of Sauchieburn. He'd been fifteen at the time, and looked twelve—nevertheless, everyone agreed even then that the young prince really had something going on. He's already had countless mistresses (including of course, two or three Margarets) and fathered five children, two of whom he acknowledges. He's been bred to be king, as purposely as a farmer breeds his finest heifer with the biggest bull. Bred, and then intellectually shaped by his tutors, dressed by royal tailors, fed the choicest produce and meat. Every day of his life so far he's been bowed to. But still—he has days when he feels a bit sorry for himself and thinks: *Did I ask to be King? I did not.* He wonders about the three previous King James-es. Maybe there's only one James, and he is simply the current physical embodiment. Not a nice feeling.

On the day he meets my Margaret in the bluebell woods, he's in a good mood. He has shoulder length dark hair topped by a black beret with a silver brooch pinned to it. His eyes are as navy blue as hers, and seem intelligent. (By the way, currently only 8% of people world-wide have blue eyes, and they all share a common ancestor. I'm talking about Percy from the Black Sea, who we are not visiting because I dislike him too much. It's disturbing to find an ancestor who is cruel to his dogs.) James's lips are thin, also like hers. Margaret is essentially a girl version of James, and he's a boy version of her. He's on horseback and surrounded by an entourage of clansmen and guards, mostly strong young men called Stewart, Campbell and Robertson. Margaret and her sisters had known the king was somewhere in the neighbourhood, visiting from Stirling, and have come to the woods under the pretence of bluebell gathering. (Think of Camelot. *It's May, it's May! The lusty month of May, when everyone goes blissfully astray!*) When a few of the men dismount, the girls curtsey—but not too low. They are, after all, the daughters of Lord Drummond. Liza Todd, who trails behind Margaret wherever she goes, also curtseys. Then James, still on horseback, indicates they can all rise and approach him.

He leans over, takes Margaret's hand and kisses it dramatically. Holds on to it. After a frozen moment, her sisters dissolve in giggles, and some of his own men laugh too. He's a ridiculously naff flirt, and although he's king and no one can say no to him, sometimes people can't help laughing at him. Just looking at him is kind of funny. To his credit, he doesn't mind. He laughs at himself too. It feels almost like a joke today, that he is king. *King!* He's not really king material. Which, of course, makes him ideal king material. The right mix of humility and confidence and humour.

'May I enquire after your name, fair maiden?' he asks, finally freeing her hand and dismounting.

'Lady Margaret Drummond, your royal highness,' she says, eyes lowered. She's thinking he's not as tall as he looked while on horseback, and his breath smells a bit like onions. Still, there's something about King James Stewart. *Those navy blue eyes.* Or it might just be the aphrodisiac effect of his title. Would her heart pound if he was, say, Jimmy the cobbler?

'Margaret! A nice enough name, I suppose, but not lovely enough for you,' he says, knowing full well how greasy he sounds. 'Way too many Margarets in the world. May I call you …Marguerite?' His eyebrows arch to show his tongue is half in his cheek.

She instantly curtsies and says cheekily: 'Bien sur, Sire.'

All the gentry know French, and quite a few of the non-gentry too. France and Scotland are best friends, while Scotland and England are like siblings forced to get along with each other. That whole dynamic pivots on this encounter, or will if Margaret has her way—but she doesn't know any of that. For now, it's just the stirring May breeze, the presence of a handsome man she hasn't kissed yet, and the sweet scent of bluebells.

'My Lady Marguerite, will you kindly walk with me a ways?'

Might as well play the king card he thinks. It's his royal duty to be perceived as highly sexed even if he is not, which he isn't. This is partially the source of his self-mockery, which ironically makes him more attractive. He is cursed! More than most people, James has two lives, two personalities. Public and private. He takes her by the hand and leads her off into a bower of a meadow. Liza Todd tries to dissuade Margaret, grabbing hold of her other hand.

'This will end in tears, Mistress! You must come away at once, think of your reputation!'

But Margaret unpeels Liza's hand and carries on. The king's men and Margaret's sisters follow discreetly, with some whispered banter between them. At one point James waves one of his hands urgently above his head without looking back, and one of his men rushes up with some rugs and a flagon of mead. James waves again, a different signal, and another man rushes forward with a basket of seed cakes and dried figs. All this is enacted with a great sense of theatre and fun. Camply. He exaggerates the romantic gestures because to do so amuses to him. Plus they conceal his utter lack of desire. He's not gay, he's just borderline asexual, which is a very un-king-like thing to be. He could take it or leave it. Perhaps his very sharp brain is to blame. Right now, even under the laughter, he cannot help thinking: *Must I really keep up this randy goat act forever? Why why why must a king always be bonking some fair maid? Oh sweet Jesus, how many more hours before I can justify reading alone in my room with my farting dog asleep by my feet?*

~~~

Nevertheless, within a fortnight, with the help of her conniving sisters and the king's men, Margaret is apparently having sex with James on a daily basis in the little room off the Drummond castle library. He's found an excuse to remain in the area, something to do with tithes and boundaries. No one reads in her family, so the library is always empty. Is it good sex, given his low sex drive and fondness of onions? Normally I look away when intimate things occur, but because of the price she'll pay, it feels important that she at least enjoys it—so I peek once or twice. Not more than seventeen times, and my conclusion is that fun is had, but not that much. More like brother and sister, a lot of talking and giggling and cuddling. The act itself is infrequent, brief, and appears to give minimal pleasure to either. Like scratching an itch quickly (or satisfying a biological imperative quickly), so they can get back to the immense pleasure of simply being with each other—which has been a surprise to them both. Friendship was the very last thing they'd expected that May day in the bluebell woods. At any rate, passionate or not, they are a successful mating pair, which is all that matters in terms of my existence. Things like orgasms and romance are icing on the cake. Though she does giggle an awful lot when he puts her toes in his mouth one day, and things improve for a little while after that.

On the upside, James is a talented linguist. He speaks Italian, Spanish, Latin, German, French, Flemish, not to mention fluent Gaelic. He's very close to his mother and his old nurse Agnes—both highly intelligent women—and maybe this predisposes him to feel comfortable in the company of intelligent women. It's only when wooing, that he becomes awkward and false—the result of no mojo. Basically, his courtship style is the laughing stock of more than one royal court—but who can say no to a king? More to the point, who would accuse him of lacking virility? Margaret knows about his previous conquests, the other Margarets, and about his children by them.

'How do I compare?' she asks sleepily one morning, her dark hair spread out on the white pillow.

'With what?'

'Your other women, silly. Am I the best?'

'Oh indeed, you are by far the most beautiful, the kindest and most cultured,' he says archly and then yawns. 'Fancy a game of cards?'

'And you love me the most?'

'Oh crikey. I do.' And it's true. In fact, he's not really loved any of those other ones. He thinks of them now as practice women. Inferior in every way.

'So will we be wed?'

'This instant, ma chérie.'

They sit up in bed, both naked, and she quickly braids three of her hair ribbons into a rope. (One of the ribbons, the red one, looks familiar to me. Again.) They hold each other's hands together, then he ties the ribbon rope around their hands using his teeth and two of his immobile fingers. She grabs the other end of the ribbons with her teeth, makes a knot using one of her immobile hands and pulls it tight. They giggle at the sight of their four hands in ribboned bondage, but then a sombreness settles over them. They stare mock-seriously into each other's very similar eyes. Their identical long dark hair and pale skin adds to the illusion they are looking into a mirror. Nakedness helps. There is no king or lady in this bed, just a boy and girl uncannily in tune with each other. There is comfort in all these things, even if there is no hot sex.

'I belong to you, James.'

'And I to you, Marguerite.'

'So say them! Say the handfast words.'

'I don't know them.'

'Jesus, *you do so*. But I'll say them, then.' She closes her eyes and says: 'By this knot, we are bound together. May the knot remain tied for as long as love shall last.'

'Is that it?'

'Sort of. There's other words too, but that's all we need for now.'

Then they kiss, which is kind of sexy with their hands tied. When the ribbons begin to loosen, they make love until it's time to part, albeit with no orgasm for either. More like a pleasurable walk that turns into a sweaty hike.

'Goodbye, wife.'

'See you later, husband.'

The unravelled ribbons streak across the sheets—yellow, red and blue. He winks at her in his self-mocking exaggerated way as he goes out the door, and this makes her chest fill up deliciously. *This, just exactly this, is what love is*, she tells herself. What began as a farce has become deep drama. They love each other.

'Ha!' she shouts.

'Ha!' he shouts back from the other side of the closed door, and she listens to his footsteps going down the stairs. Imagines him taking a first breath of morning air and striding over to his horse and waking his man, who'll have been patiently waiting all night. Then she buries herself deep under the sheets and thinks she has never been so happy.

~~~

My grandmother times 15 has been made, and six weeks later Margaret breaks the news excitedly. After all, James is unmarried, and her pregnancy will prove she's a fertile woman worthy of wifedom and also queen-dom. My God, she can't wait to get her hands on the grand hall at Stirling Castle. Add a lick of paint and get rid of those dusty old tapestries. She's made a list which includes items like *Ban dogs from ballroom, new rugs in bedrooms, get rid of cracked china.*

'You're what?' he says, pulling away from her.

'I'm with child. Your heir.' She proudly pats her flat stomach. Her dress front is open to the waist.

'Margaret,' he says, in a new voice.

'You mean Marguerite, right?'

'Marguerite, this does not bode well, my deerling.'

'Why not?' She laughs nervously, thinking he must be kidding. James has a weird sense of humour sometimes.

Long pause. A dog is barking somewhere.

'I'm betrothed to Margaret Tudor of England. It was arranged a decade ago. Before I even knew you.' He says this in a rush of air, while looking miserably at the floor. 'I'm sure I told you this. *I know I did!* It's politics, not love. I didn't feel there was any deceit or even betrayal in what we have. Knowing Margaret T will be my queen one day.'

'Oh, her!' scoffs Margaret. 'You don't even know her! You'll probably hate each other.'

James sighs and looks older.

'That's irrelevant. If I don't marry her, England will be meaner to Scotland. Our families have agreed.'

'Oh pooh! Families schmamilies,' she says, but her bravado is false, and he knows it.

Three minutes pass, while they turn from each other and begin to dress.

'I don't get to marry for love, my love. I thought you understood that.'

'But, but, but we handfasted.'

Then he shakes his head, sighs and leaves the room. Margaret stands there with her freezing blue nipples and crying blue eyes, buttons her gown up and shouts:

'Liza! Liza! *Oh Liza!*'

By the time Liza finds her mistress, Margaret is on the floor in a paroxysm of weeping.

'But I'm having his baby!' she sobs. 'He has to marry me!'

'Well, he doesn't actually,' says Liza, on the floor too now, rubbing Margaret's narrow back. She lets her cry it out, which takes a long time—almost two hours. At the end, Margaret's face is shiny with snot and her eyes are red rimmed. She's hiccupping and saying things like:

'Now no one's ever going to call me Marguerite again!'

And:

'I hate him! I hate him! I hate him! By God's bones, he's a sard-face!'

(Sard meaning fuck. Sard you! Get sarded! Away to sard!)

'Yes dear, yes my sweetling. He certainly is a silly sard-face.'

Liza likes James, she even feels sorry for him—but she keeps saying things like this in her sing song voice, while stroking Margaret's hair and shoulders and cheeks. She refrains from saying *I told you so*.

~~~

The child is born on a snowy March morning, and called Margaret of course. Her last name is Stewart because that's the truth. Too many people know she belongs to the king, and in any case, he's already promised to claim her as his illegitimate daughter. (Quite a common thing then, and not as scandalous as you might think. In many ways our morals are more prudish and hypocritical.) There's been no wind, and outside the castle snow sticks to everything. When the sun comes out, the whole place sparkles. Liza stands at the window and appreciates it. Says thankyou to God for the safe delivery, but Margaret the mother keeps to her dark fetid bedroom for weeks, and she curses her situation. The maid who keeps the fire going, and the wet nurse with the new baby at her breast—they are no comfort to Margaret. Her sisters are afraid for her, and her parents are angry with her—none visit, aside from once to see the child. Not even Liza can help, for now the child is here, Margaret has no purpose, no value, no importance in the world.

~~~

The two Margarets come to visit James six months later. It's not easy, sneaking into the castle of the king of Scotland after you've been banned by decree, but Margaret has never let things like that stand in her way. She fibs to Liza, says she's going out because she heard marvellous things about a new dress-maker near Stirling.

'How did you get in here?' James demands in the dim corridor where she finds him.

'Magicked myself here.'

'Oh Margaret, please. You'd better leave, and take the wain with you.' He tries to walk past her, but she blocks the way.

'The wain? The wain?' She thrusts the baby towards him, and says: 'Do you not recognise your own daughter?'

'Please stop this foolishness,' he hisses. 'You've changed. You never used to act like this. Or look like….like this.'

'Really? I have no idea why.'

For a full thirty seconds they freeze and glare at each other. Then the baby burps and smiles at him. He smiles back. 'Don't use the child like this,' he says in a softer voice. 'She's innocent and yes, I own that she's mine. She has my name and I support her, do I not? Do you not receive the money?'

'Aye, but the money's for her, and I, I, I miss you. And you must miss me too, do you not?'

She hasn't meant to say any of this. He grabs both her upper arms and softly shakes her. Tears tremble down her face.

'By Christ's fingernails, I miss you every day, Marguerite.'

Then they kiss, a mad mash of a kiss till their lips feel bruised and swollen. Annoyingly and inconveniently, this is the hottest kiss of their relationship. Little Margaret cries out because she's being squashed between them, and they pull apart.

'Then marry me instead, not silly old Margaret Tudor,' she hisses. Some of her spit lands on his face.

In the distance someone drops a glass, someone else yells that water is boiling now and *What the hell will I do with it?*

'Hello there, young Margaret, look at you!' He takes his daughter into his arms. 'Pretty little thing that you are, you are, you are.'

Margaret the mother leans sideways against James, while he croons. Feels herself pour into him, latch on to him, as if she's found land at last after countless days drifting at sea. Oh, the sheer bliss of her mind emptying out! The baby keeps smiling.

'Listen, Margaret,' he whispers to his daughter. 'I'll fix it somehow. I don't know how, but I will. Nothing makes sense otherwise. I'll talk to people. There's got to be some perks to being king. I mean, if I can't choose my own wife and raise my own daughter, then what's it all for?' Then he looks straight at Margaret the mother.

'Marguerite, you've got to be patient and not tell anyone, alright?'

'Oh, *thank you*,' she whispers.

'Come to the feast a week next Saturday. Bring your sisters.' He's grinning wildly now. 'Bring our beautiful daughter.'

~~~

Margaret and her sisters arrive triumphantly at the castle in their finery, silk gowns of blue, lavender and yellow. Liza is with them too, in a black dress, holding baby Margaret. They are all greeted warmly, seated at one of the bottom tables and given venison and barley cakes and beer and wine. The men who need Scotland to align with England are especially solicitous, but Margaret doesn't notice. Politics bores her, and she takes their friendliness at face value. Or maybe their friendliness is genuine, and her position is so unthreatening they hardly notice her. Either way, within an hour, Margaret and her sisters are lying on the floor dead, three fallen flowers. *Poisoned,* says the castle doctor shaking his head, though whether it is food poisoning or something more sinister is unknown. No one else from the feast is sick. Not even Liza, who has eaten the same food.

~~~

The day after the funeral at Dunblane Cathedral—three beautiful young women cold and white in their open caskets, the place heaving with mourners—James sends his baby daughter with Liza to Edinburgh Castle. Nothing terrible will happen to young Margaret Stewart. Or to her children or their children. (Isobel Ogilvie, her great granddaughter, will be my 11th great grandmother and live in Fairburn Keep, ten miles from where I sit writing this book.) Under the King's protection, even after he is dead they will thrive and multiply and mostly live long lives. It's only poor Margaret Drummond with her white skin and navy blue eyes who gets a rough deal. Is she dead from accidental food poisoning? Or is she murdered as an act of political expediency? And—more pertinently—if she's murdered, is it with James's knowledge or even consent? I can tell you this: For the rest of his life, until he dies aged forty, he pays for weekly Masses to be said for her soul.

~~~

Yikes! Just like that, we're back to the first birthday.

'Look what I've got for you, Margaret.' I pull out a small gift from my robe pocket.

'What is it?' She wriggles out of Liza's arms, and slides down her body until she's on the floor again.

'Off you go, then pet,' says Liza who starts squeezing a pimple, proving I'm invisible to her. Good. It would be awkward if the grownups saw me. They'd probably want to make small talk, and I've never been good at that. Plus, talking to anyone is exhausting.

'Here you go.' I give Margaret the present.

'Oh!'

'You're disappointed?'

'It's okay. It's just I already have one of these.'

'Ah. So spinning tops are a thing now?'

'They've been a thing since my grandmother was a baby.'

I rustle around in my robe pocket for something more exciting, but all I can find are scrunched up tissues, some old pennies and a paperclip I rescued from the hoover yesterday. I give her the pennies and paperclip, then remember I've got a red ribbon tying my hair back. I quickly loosen it and hand it over.

'Who are you, anyway?'

'Your 15th great granddaughter. I'm one of the things that evolves from you.'

Silent pause, while she looks at me expectantly. Also with utter bewilderment.

'You're actually part of a really big crowd, my 15th great grandparents,' I assure her. 'There's 262,142 of you. But you're the only one I know the name of.'

More silence, so I go for it.

'And there's an even bigger crowd of your descendants. Roughly 172,186,884 of us. Check us out.' I do a flirty whirl, ending with a grand gesture to indicate the lucky hordes—each one a result of Margaret saying yes to the king of Scotland when Liza Todd wanted her to say no.

'That's so funny,' she finally says, in a tone that says I am not funny. 'You're such a liar.'

'Yeah, you're right,' I say. 'Busted.'

~~~

I'm lying about being a liar. More of a social fib, because no one likes to be corrected. This is a true story, but there's a time and place for truth. Even if I told her about her destiny, maybe she'd still feel lucky. After all, true loves don't grow on trees, and many people go to their deaths having never known what kisses can be. James's destiny isn't so great either. Margaret Tudor is hard work, even for a king.

Jeanne Breconnier

1652
Paris, France

I find it a little disorienting to be in France. In my family, we're proud of our Irish and Italian ancestors and pretend we've no Germanic blood (we've probably 15%)—probably an unconscious post-war bias—or any other ancestry. Until recently, I'd never even heard of a Scottish connection, so visiting Margaret Drummond was a surprise too. It's a strange vanity, to be prouder of Irish and Italian DNA, given that in those early days of emigration to California, they were near the bottom of the ladder, a millimetre above native Americans, Chinese and Africans. The literate and wealthier Germans and English immigrants ruled San Francisco society, while the French immigrants mainly stuck to the east side of the continent, mostly the far north of what becomes Canada. (We'll get to that place soon.) But nationalities come and go, in a fashionable sense. There is a tendency in my family to be prouder of nationalities that have not conquered too many countries or initiated genocides. Or maybe it boils down to something more basic. The food in Italy is hard to beat, and no one is funnier or better at writing than the Irish. Stereotyping may be wrong, summing up any group of people based on their ethnicity is simplistic and offensive, but have you ever had a bad meal in Italy? Or spent time with Irish people and not laughed? Besides, we couldn't stop stereotyping if we tried. It's wrong, but it's an automatic reflex. Making sense of the world in a time-efficient way depends on it. Oh, wait! That's just a defence of positive stereotyping. The negative variety is, without exception, insulting and inaccurate.

Harmful and dangerous. That reflex should be nipped in the bud every time it rises up till it dies. Even if, especially if, everyone around you is making the same ugly assumptions about a group of people. Remember, they are all your fellow Homo sapiens!

~~~

As you've no doubt noticed, as I get closer to my own time, I'm visiting the past more frequently and staying longer. This is not just because I'm more interested in my own species and times, although I am of course ethnocentric. We all are, since we cannot help thinking of ourselves as the centre of the universe—at the least, we cannot know any other perspective. Even Kevin was ethnocentric. It may be the only unchanging maxim of existence. But I'm also stopping more because Life is evolving more quickly. Remember the clockface with all of Earth's time up to my existence plotted into sixty minutes? Kevin appeared at four minutes past the hour, but—incredibly—the current birthday girl doesn't appear until the tiniest millisecond before twelve—and yet, the changes between her time and mine will be far more fundamental than changes that took billions of years to occur. Some days it feels like we've become a compressed bundle of restless DNA hurtling towards…I don't know what, but I hope it's good. Or at least not horrible.

~~~

Today we're half way through the seventeenth century, smack dab in the middle of Paris. This is the 11th arrondissement, one city block from the Seine. It's called Popincourt and has more people per square inch than any other European city. It's cold and dark because it's January, and anyway—this room has only one window, that narrow slit looking out to the courtyard, where people hang their washing even in winter. That's my 8th great grandmother, Jeanne, over there sitting up in the open bottom drawer of a tall dark chest of drawers. She's descended from poor

Margaret Drummond in Scotland, and will eventually end up in me via Richard Swarts, who intersects my teenage grandmother's life for three years. Long enough to impregnate her twice before hopping on a freight train never to be seen again.

But back to Jeanne.

The open bottom drawer is her bed. She's got pretty hazel eyes like my mother, but she's not very healthy or clean looking, is she? Bit of a runt.

'Happy birthday, ma chere,' I say.

'I'm not your chere. Dear, oui. Yours, non.'

'Sorry. It's practically the only French phrase I know.'

'I understand English, moron.'

I think I know why she's being cranky, aside from the cranky gene. She's had a rough start. (Biological imperative of crankiness and impatience? More efficient use of time and resources, less susceptibility to peer pressure and less tolerance of unacceptable situations. Cranky people don't give a shit. Downside? Fewer friends. Also the feeling of crankiness, which itself can increase blood pressure and make you crankier, *god dammit*.)

'Jeanne. I wanted to say I'm sorry about the incident leading to your birth. But truly, there is no shame attached to you. You're innocent, it's not your fault.'

Two minutes' silence follows.

'What the hell are you talking about?' She has a French accent a la Juliette Binoche. As sexy as a baby is allowed to be. No matter what she is saying, she sounds intelligent and classy.

Jeanne is the product of incest. She's not my first grandparent to be born from inappropriate intercourse. So far, sixty-four have been the product of rape and 238 from incest—and yes, much of the incest could be counted as rape, but not all. Not by a long shot. Her uncle had sex with her mother, his thirteen-year-old sister. It's taboo but not as dramatic as it sounds, although it was not a pleasant experience for her mother at the time. (She said *non!* so it can be classified as rape.) Along with thirty-two others, those two people still live in this house. They're not bosom buddies, but they get along well enough, considering. Their offspring, Jeanne, is being raised with her aunts and uncles as if they are her siblings. She calls her grandmother *Maman*. In less than a year, another one of her uncles will be born and take over this bottom drawer. It will be seven years before it reverts to holding clothes again, and the clothes stored there will always smell of pee and breast milk.

The reality of rape is never good news for the female—but then so much is not good news for females. From the time of the biblical Eve and the mitochondrial Eve, women were both revered and—perhaps not ironically—feared. They were physically weaker, but had this advantage over the male: *They always knew who their offspring were.* Children—sons to protect you and plough your fields, and daughters to barter for breeding purposes and to expand your tribe—meant power. Men (even married men), until recent DNA testing, could not know with certainty who their children were. They had to take the woman's word for it. This was a major inequality, which may or may not be the reason that from early times men wanted to own and control women. Fathers owned daughters until they sold them, or dowry-ed them, to husbands. Husbands owned wives to do what they liked with. An unmarried woman, up until my childhood and maybe continuing still, was considered a failure and a pariah. One of the popular card games in the 1960s was Old Maid. She had a wart on her huge nose and wore glasses. No one wanted to be stuck holding the old maid card. Old maids had cooties and lurgies, obviously.

Human relationships are never straightforward, and of course not all men want to own and exploit women – consciously or unconsciously. We are all, to an extent, victims of our natures and histories and of societal expectations. Lots of men look up to women, honour them, respect them. But the bottom line is we are all products of sporadic historic rape and incest. I find it disturbing, so I avoid first birthday visits associated with abuse of any kind. I mention it here only to remind you that we owe our lives to the occasional success of wickedness. If you must take pride in your family history, do it with humility, accepting that we are all results of some people being horrible to some other people. Kindness, morality, fairness, equality—perhaps these only become possible when the battles and natural catastrophes are over and a species is fully established.

~~~

In evolutionary terms, what are the biological imperatives of rape and incest? Are they part of the compulsion to make babies, or are they something more sinister? Can such moral transgressions ever have an evolutionary justification? I can see incest being inevitable and necessary in small remote communities simply because everyone around would've been a relative. Incest even in that context is largely taboo now, and theoretically rape is too. And yet, rape remains an unsurprising, albeit disgusting, consequence of—for instance—soldiers invading enemy territory. Raping might be hardwired into the male brain (yes, even the brains inside really nice boys) to be triggered in certain situations. To literally plant their seed in enemy territory. To recreate themselves as a gesture of power and dilute indigenous genes. Though even in peacetime, there have been periods of history during which rape has been incorporated into society's hierarchies. Think of the primae noctis ritual, in which medieval lords were entitled to sleep with serf brides on their wedding nights before their husbands did. And rape as an apparatus of dominance is well documented in the non-Homo sapiens kingdom too, where it is never considered morally questionable. What would be the point of a doe, for example, objecting to being mounted by the stag who nearly died for exclusive reproductive rights over the herd, when he is her only chance of becoming a mother?

~~~

'Are you going to answer, or just keep staring at me like a simpleton?' asks baby Jeanne. If she could stand, she would have her hands on her hips. 'What exactly is not my fault?'

'Sorry. Forget it. There's nothing to worry about.'

'Merde.'

'That's a swear,' I say, proud to prove I know another French word, even though I'm not sure what it means.

'Tu dis de la merde !'

'Okay. Whatever. C'est la vie. That's French too. So there!' I'm regressing into adolescence now. Not a proud moment.

'Quelle vie de merde!'

~~~

Since Margaret Drummond's time, the most obvious shape of evolution is societal. Jeanne is born at the beginning of the European spread westwards. British, Belgian, Danish, Portuguese, French, Spanish, German, Italian and Dutch empires are blooming all over the world. This is not quite the same as the impulse that drove hungry hominid Freddie northwards from Africa, no no no! That nomadic gene sprouted wings and is now driven by ambition. Put simply, Homo sapiens have a burning need to own land, probably because land is power. Stealing it is morally justifiable because beings who are weak enough to surrender don't deserve land. Yes, of course we are still grappling with the consequences. And worse, we are perpetuating the same impulse in more insidious ways. Empire-building has probably been in our genes since Jeff's day, but in Jeanne's time it rises to the fore.

~~~

Let's leave this scrawny sweary child for a while. Jeanne becomes a King's Daughter in eighteen years. *Un fille du Roi.* Which is the solution France comes up with when their colony in New France begins to falter due to lack of women. By the time Jeanne signs up for the adventure of marrying a man she hasn't met who lives 3,419 miles away, the programme has been running for a decade with over 800 French girls now married and breeding in New France. In 1673, Governor Frontenac of Quebec places an order for sixty brides-to-be. They are not matched to specific men beforehand, but their obligation to marry is part of the deal. My Jeanne is given a dowry of clothes and money from the King, plus her passage. She kisses her family goodbye—the ones she's still speaking to—and boards *L'Nativite* at La Rochelle. She waves goodbye cheerfully, not minding at all she will never see them again.

'Salut les loosers!' *So long, losers!*

It's not a bad trip, considering the foul food and hard hammocks and lack of privacy and the seasickness, to which Jeanne is susceptible. Most of her fellow passengers are orphans or girls like herself from the slums, for who else has nothing to lose? Who else would jettison everyone and every place they know for the complete unknown? The main advantage of a poverty-stricken childhood is low expectations of comfort, and Jeanne arrives in Quebec city on September 3rd whistling le Carillon de Vendome. The second night she goes to a dance in the town hall organised for the local bachelors to meet the King's Daughters. She wears the yellow silk gown she's been given for the occasion. It's a little wrinkled from the journey, a little mouldy smelling, a little loose, for she's lost some weight—but youth carries off imperfections. She has powdered her face and rouged her lips, soaked her wrists in lavender water, and around her neck is a thin gold chain given to her by her mother/grandmother the night before departure. (*Je t'aime ma puce. Bon voyage!*)

Tonight the hall has over two hundred men and fifty-eight King's Daughters, two of them having died at sea. The women are wearing identical silk gowns with satin ribbons under their pushed-up bosoms. The dresses are pastel colours, and it's like a spring garden has been sprinkled into the men's murky midst. Roses, primroses, violets, bluebells and pale ivy, all trailing the nostalgic scent of lavender. The men are mostly in shiny black suits, their hair and faces and hands newly scrubbed. They seem shy of the women at first. Standing in small groups, swallowing their coarse words at the back of the hall near the beer table. The girls walk around, arms linked, giggling and chatting, peeking at the men behind hands and fans. It makes the men feel strange. A few are suddenly homesick enough to think of their mothers. The music starts—two fiddles and a squeeze box. The girls instinctively line up, their backs to the wall, and face the men, who instinctively line up in front of them, hungry for touch. It's quite a heady experience for Jeanne, to attract so much attention. She's not unattractive—she has beautiful brown-green eyes—but nothing like this would ever happen to her in Paris. She's too tall, too thin, too clumsy, and it's been pointed out to her that her chin is a mite weak. She feels drunk on the attention, and this has the effect of making her pretty. Cheeks tightened from smiling, and pink. Lips moist and merry. Even her movements become coquettish and graceful.

Thirteen men propose marriage to Jeanne that first night. One of them is a fat little man called Charles Etaline. He's no longer young, though his fatness smooths out his skin. *Non, merci!* she says to them all, and Charles gets a kiss because he looks so sad. There is something wrong with each of them— too short, too ugly, too scarred/crippled, too old, or too fat (in Charles's case). But towards the end of the evening, she dances with Louis Chapacou. He is the right height, age, the right amount of handsomeness and health. Because of these visible advantages, she credits him with intelligence, kindness, loquacity.

The band is playing a slow sweet waltz, and he dips his mouth close to her ear.

'Épouse-moi, je suis déjà riche. Je vais te rendre heureuse.' *Marry me. I am already rich. Besides, I will make you smile.*

'D'accord!' exclaims my Jeanne, then throws her head back and laughs like she's being tickled. Can life get any better?

Louis smoothly produces a ring from his pocket, kneels down, slips it on her finger. This scene is happening all over the hall, but the dancers nearby stop dancing to cheer. *Bravo!*

They are married on September 10th, her fifth day in New France. But Louis is not rich, not by a long shot. The son of settlers, he lives with his parents in a log cabin in the middle of the woods where grizzly bears roam. Bears! She'd heard about bears in Canada of course, but somehow not believed in them. And more worrying, Louis cannot seem to do the deed. Jeanne is not a virgin—a half-brother took care of that years ago, continuing the family tradition—but even so, she expects her husband to take the lead. Louis simply yawns every night, ignores her sexy nightie, kisses her with his mouth closed before turning away and saying:

'Je suis si fatigué. Bonne nuit ma chérie.' *I am so tired, goodnight my dear.*

This will not do, thinks Jeanne, and runs back to the Quebec boarding house before dawn. The marriage is annulled for non-consummation that day, September 15th. Louis, who is gay, will marry again in five years and have better luck with an older widow who is delighted not to have intercourse. She has seven children and is exhausted. He's delighted to raise step-children, because not having children was going to be the biggest regret of his life. This will marginally cancel out the life-long suppression of his true self.

~~~

Meanwhile, my Jeanne is back to shopping for a husband. It feels like she's been in New France for years, but it's only been a few weeks since she disembarked from *L'Nativite*. Most of the girls she arrived with have gone now, off into their shiny new marriages. The ones that are left encourage her to take her time, to wait till the right man comes, but Jeanne has never been patient. Three days later, on September 18th, she's at the altar again. Crespin Thuillier dit LaTour is a young widower from Boucherville near Montreal. His first wife died recently, and he's handsome, in a dark, sturdy, no-nonsense way—like a solid pony. He's still grieving, but a man needs a wife in this country and he's delighted with Jeanne. Such a practical young woman and his hands meet around her waist, she's that slender. She moves to his house outside Boucherville, and his affection and sexual prowess cancel out her dread of bears. All in all, she's whistling a lot, and doing little dances around the cabin when he's out hunting or ploughing. Je suis amoureuse! *I'm in love!* She gets pregnant immediately, but her son dies just after Christmas, aged six months. She doesn't know why, he simply doesn't wake up one day. It's unbearable. She wraps his body in a lace shawl she's brought from Paris, and rocks him close to her chest for hours. Eventually her husband gently unpeels her hands and removes his son. He's buried in their woods with a tiny wooden cross above him, next to the grave of the woman whose place Jeanne took. The first wife.

It's a bleak time. She needs another baby—they both do—and by Easter she's pregnant again. She cooks Crispen breakfast one morning—bread made from ground hazelnuts, fried rabbit thighs, and a cup of coffee (their only luxury). She kisses him goodbye. He kisses her back and then kisses her belly. He starts to rub her nipples through the fabric of her dress, and for a moment she closes her eyes and leans into him. Crispen always has this power over her, it's like a magic trick. He touches her breasts, all thoughts disappear, and oh! Isn't it bliss to not think? But then he withdraws his hand and says:

'Non, pas maintenant mon amour. Plus tard, plus tard!'

She kisses him one last time, thinking he feels a little feverish—he's been complaining for days now of feeling tired, maybe he's sick? But off he goes, regardless. That evening, his body is returned to her strapped to his horse, which is led by a young Anishinaabeg man who does not speak French or English. He is wordless but with sign language explains he found Crispen dead on the mountainside. The body has no marks, no wounds, aside from the cut on his hand from a few weeks ago. The cut is still red and swollen. It festers as if he's still alive and it smells slightly. She knows things like tetanus can kill a person but she still can't believe it. Her baby kicks for the first time, but she doesn't notice. Nor does she cry. The young man buries her husband where she tells him to, next to their son and his first wife. Now there are three little wooden crosses, and it's just her in the middle of nowhere with a baby curled up inside her. Jeanne packs what she can carry and goes back into Boucherville on a horse. It takes almost three hours, and all the way she allows herself to holler and cry—so that when she arrives, she is emptied out and dead-eyed.

~~~

Life is no longer an exciting adventure, and being alone is just lonely. For a long while her pregnancy doesn't show. She works as a prostitute servicing the lonely French immigrants who've not been lucky enough to get a King's Daughter or similar. She doesn't apply the word prostitute to herself, naturally. These are, in general, gentle shy men who are so grateful, they leave her a gift of money on her bedside table. Isn't that sweet? Each one is a possible paramour, she tells herself. The amount of money they leave is never discussed, so definitely not prostitution! After each man leaves, she washes and takes out her rosary to say ten Hail Marys, more if she has time. She's Catholic—everyone here is. Once a week, she scurries to the dark confessional of St Jerome's. *Bless you again, my child*, says the priest, and each time she feels her sins fall away.

~~~

Homesickness arrives at last, now her defences are down. It makes itself right at home in her chest, throat, eyes, guts—a chronic heaviness, an unanswerable yearning. And her thoughts are increasingly backward-looking ones, until Paris is sweet-scented and painted in sentimental colours. There can be no cure, for this Paris doesn't even exist. Homesickness is a sickness of the soul. Maybe a sickness so strong, so tenacious, it seeps into our DNA epigenetically and seeks a remedy way beyond our lifespan. When I was young, I used to travel. I'd go to Scotland, Ireland, France, Italy—places where I knew no one—and sometimes I'd experience a sensation of coming home. It would fill me up, and I'd sit somewhere and be quiet. Not let the homecoming joy spill over. Maybe, in some hidden part of me, curled up inside helixes of DNA a thousand miles long, my ancestors were doing cartwheels. My Irish immigrant ancestors beside themselves because I happened to be in the town they grew up in. I remember when I got to northern Italy, having hitchhiked through France, I felt a strange whooshing sensation. My ears popped and my chest filled with light—maybe my Domodossola ancestors were breathing out a sigh of relief. Again, in Paris. The same calm inexplicable sense of rightness when I strolled through Popincourt, a block from the Seine (no longer stinky). Not all the time while travelling, and never while sight-seeing or ordering escargot, but often while sitting in a public park munching on some dry bread and cheese. A sense of familiarity and ease. *But why wouldn't you feel happy here? Home at last!* my immigrant ancestors might've been saying. Jeanne might've been saying. *Look, look! Look how Popincourt's changed. Oh, but it's the same place under it all.* Of course, I didn't hear those voices. I didn't know I had long-ago family in those places. I was too young and self-centred to even care. I filled the world up entirely, neatly, and the past and future were unreal. They did not contain marvellous me.

~~~

Jeanne's marriageable value has shrunk considerably. She's now twenty-three years old, twice married, and whether she likes to use the word prostitute or not, sleeping with lots of men does something to a girl's reputation. Then along comes a customer called Charles Etaline. Jeanne vaguely recognises Charles from that heady night of thirteen proposals on her second day in Canada. *He's put on more weight*, she thinks. *Maybe best if I go on top.* He recognises her too, but neither say anything about it. In fact, they manage to not say anything at all. He's clean, anyway, and that's always something—she says to herself. The sheets look clean, anyway, and that's always something—he says to himself. After the sex (perfunctory) she rises to wash and dress. He lies in bed watching her by candlelight, noticing her slight belly bulge, her swollen breasts. He drops some coins on the bedside table, gets up and dresses, turns to go. Then grunts with his back to her:

'Je pense que nous devrions nous marier finalement.' *Reckon we should marry after all, then.*

She's struck by how familiar his accent is, now his words are strung together. Even within Paris, the accents differ enormously according to district.

'Vous n'êtes pas de Popincourt, n'est-ce pas ?' *You're not from Popincourt, are you?*

'Bien sur que si.'

'Ah! D'accord.'

This makes him more acceptable and she smiles warmly. They were made in the same place, of some of the same ingredients. It's surprising how much safety stems from this. And surprising how much safety matters, now that her expectations have shrunk. They marry that Saturday in a suburb of Montreal. She wears the yellow gown, which is now faded but amazingly still fits over her pregnancy. (*Thank you King Louis XIV.*) It turns out Charles is quite a catch. Renowned as a shoemaker, he's also a farmer with his own patch of land at Longueuil. The evening of their marriage, he takes her home. It's only a mile away along the St Lawrence river, a little house made of field stone with a thatched roof. Aside from the vows, he's hardly spoken. Jeanne has been quiet too, for hasn't she just surrendered all her dreams of passion and romance for safety? It's wise, but also a little sobering. When she imagines the night ahead, her spirit shrinks. But the house cheers her. The kitchen is well equipped and the roof does not leak. She braces herself for her wedding night. Like a chore to be done, a duty. In the darkened bedroom, he strokes her hair and whispers: 'Do you want me?'

'Bien sur! Oui!' After all, she's had lots of practice pretending to like sex, and they've already done it a few times.

'Really?'

Pause. His hand tenderly lifts her face until she must meet his eyes. Which are dark and honest.

'No.'

'Ah. I thought not.'

'But I want to want you,' she says in a rush, because it's suddenly true. 'Does that count?'

'Maybe. I don't think so, but maybe.'

It would be nice to say Charles then moves to the spare room until a day hence, in about three months, when she signals subtly and romantically that a spousal visit that night will not be unwelcome. What happens instead is they have unmemorable wedding night sex, which is followed by months of the same.

At least he finishes quickly, she tells herself each time.

~~~

In January she gives birth to a girl, and they call her Jeanne Thuillier dit LaTour. Charles will raise her, but not give her his name—and no one expects him to. A year later Jeanne gives birth to her third child, and first by Charles. Catherine-Teresa is my 7th grandmother, and a good sleeper from the start, which is lucky because Jeanne needs all her energy to adapt to farm life. She's quite squeamish about chasing pigs and milking goats and killing cockerels. But she's smart, my Jeanne, and learning takes her mind off her troubled past. Soon she's managing the animals as well weeding the vegetable patch as quick as any country girl. And she churns the creamiest butter her husband has ever tasted, or so he says. Adaptability being a primary biological imperative, perhaps it's not surprising she's gone from Parisian slum rat to Canadian milkmaid so completely. (Think you know who you are, what your limits are? You do not know, not until you are tested by change and your need to stay alive. We are all, to different degrees, capable of living different lives and being different people.)

~~~

Time shuffles along and she's too busy to notice. She doesn't often argue with Charles, but she sets herself against him with a tapped down habitual resentment. When she's twenty-seven, with four children and a goat herd to manage, Jeanne falls in love with Stefan, the man who collects the goat's milk every week. He's very good at talking, very astute and funny and kind. Plus he's over six feet tall (parents from Norway) and blond. His body is perfectly proportioned and very masculine, so any proximity to him makes her feel feminine and petite. Just to feel girlish and sexy for a change! Not to tower over a chubby man! Stefan is, in essence, the anti-Charles. Because Stefan is unaware of her feelings, she's able to sustain her crush for a few years. It feels harmless, loving him, and after all she has to love someone, doesn't she? What else would she do with her bottled up romantic heart, but pour it into a man she can't have?

Charles does not notice—or if he does, says nothing. He loves his food and his children and his work. He loves his life like it's a marvellous object outside himself which he happened to stumble upon and was smart enough to grasp. He wakes up each morning looking forward to the day—but he still does not see the need for conversation, and he's grown even fatter. Everything that is wrong with Jeanne's life—loneliness, sexual disappointment—would be remedied if only she was with Stefan instead of Charles. In a rash full-moon moment, she whispers her crush to Stefan and he responds with a long deep kiss. Within twenty-four hours she has jettisoned her life and run away for the sake of True Love, though she tells Charles she's going to visit a sick friend and will be back Monday. She does not burn her bridge, which is lucky, because by Sunday morning she realises her terrible mistake. Stefan is not at all who she thought he was. He is not kind, he might even be a little cruel. He's not a great conversationalist either—his stories almost always involve his being smarter/faster/stronger than someone else. Like many handsome men, he's never bothered to learn the niceties of making love, and is selfish to the point of ineffectiveness in bed. The kiss, it turns out, was the only erotic moment of their brief liaison, and she scurries back to her husband and children. Monday morning she tells Charles she thinks they'd get a better deal selling their milk to another merchant she's heard good things about.

Then my Jeanne takes a good hard look at Charles and consciously begins to grow a kind of love for him, like planting seeds which would certainly come to nothing if she didn't pamper them with compost and water and daily inspection. It's not a heady love, not heart-pounding, more of the hardy variety. No expectations of intimate rapport or passion. This new love settles into her as if it has physical weight—not oppressive, but calming, soothing. When he performs small acts of awkward kindness, they don't repulse her now. They make her want to put her arms around him as far she can (he's too wide to reach all the way around), and just hold him that way for a while, rocking a bit. They have four more children, all of whom survive to adulthood, and life is good for a long time. Whenever she thinks of Stefan, she thinks: *Whew! That was a close call!*

The years shunt by faster, as change slows. Whenever Jeanne feels the vestiges of restlessness, she whitewashes a room. (No one but the gentry can afford coloured paint.) Whitewashing gets rid of the smoke stains and grubby hand marks, but mainly it gives the needed illusion of a new place. Charles and the children become used to her impulsive whitewashing. They quite like it. It means Jeanne is in a better mood for a while.

~~~

One morning when she's sixty years old, Charles brings her a coffee in bed because last night she'd complained of exhaustion and a tight chest.

'Tiens ma chérie,' he says softly. 'Comment tu te sens aujourd'hui?'

She lies on her back, perfectly peaceful, but he cannot rouse her. The doctor is called and he sends her to the hospital in

Montreal. There's no obvious diagnosis. Has she simply run out of steam? She begins to fade, and her children and grandchildren gather around her bed in shifts. Charles only leaves her side to eat.

'Peut-elle m'entendre?' he asks the doctor.

'There's no way to know. Talk to her as if she can hear you, Charles.'

So he does. He talks like he's never talked before. He tells her how she broke his heart that first night of the King's Daughters Dance, when she turned down his proposal. That her weak chin and boyish figure and sparky nature drew him like a moth to a flame. (Weak chins indicate higher than average oestrogen levels, hence better fertility.) That the surprising sight of her in that brothel bedroom two years later, the sight of her pregnancy, nearly broke his heart again. That their wedding day was the second happiest day of his life. That he always loved her but knew it wasn't mutual. Not at first. Not for almost two decades. That he knew, of course he did, that she was in love with that Stefan fellow for a spell—that nearly broke his heart a third time. That when she finally turned to him, when she began to love him properly, it was happiest time of his life, for he'd been poised to receive her love since the day he laid eyes on her.

Charles Etaline is finally talking, but does she hear any of it? I don't know, but I can tell you this: When he's talking and holding her hand, she returns each of his squeezes. When he needs to eat and lets her hand go, a little spasm passes over her face, as if the tiny muscles used for crying are twitching. She gets agitated until he returns, sits by her side and takes her hand in his again. Jeanne Breconnier dies on 20 February, 1711 at 4:12 a.m. She's alone, for conscious or not, her heart has chosen to spend its final beats when no one is in the room.

~~~

My 7th great grandmother Catherine-Teresa brings her five children to the funeral, and later has a little snoop among her mother's things. Her sisters are there too, opening drawers and cupboards. Weeping a little, but mostly on the look-out for something pretty or useful for their own lives. After all, by the standards of their time, their mother has died an old woman. The yellow silk gown is discovered and quickly discarded. It's discoloured and in disrepair.

'Oh, I'll have it,' says Catherine-Teresa, picking it up from the floor. 'I'll make it down into play dresses for the girls.'

Charles, who will die in five months, is the only one in the house who is inconsolable, who cannot eat or sleep. For his skinny tall weak-chinned Jeanne is dead, dead, dead. And no one else will put their arms around him as far as they can go, squeeze him close and rock back and forth—almost but not quite like a dance.

~~~

And yet.

And yet, here she still is. An almost new child in a freezing dark house in Paris. She's still sitting there looking straight at me. You'd think she'd just climb out of the drawer and crawl over. But maybe she needs to conserve energy. Or maybe she can't crawl yet. One thing's for sure, she's not afraid of me. Look at that little face—it's fierce! She stinks of poop and sour milk, and Paris itself pongs. Sewage, horse manure, wet clothes, mouldy walls, unwashed bodies. Don't breathe too deep.

'I brought you a present, Jeanne.'

'What?' Suspiciously.

I give her a small parcel wrapped in white tissue and tied with a gold ribbon. She tears it open, but it's hard to see her expression because there's so little light in this room. Still, she gives a squeal, which I find gratifying.

'Ooooh!'
'Do you like it?'
'Oh non, je n'aime pas. Je l'adore!'   *Hell no. I love it!*
Who would have thought a tiny doll would bring such pleasure? It's quite life-like, despite being made of wood with a porcelain head. The blonde hair is human, and it matches the yellow silk dress. Painted dark eyes, rosy cheeks and a red rosebud mouth.

'She's so pretty,' she whispers, lightly touching the doll's face.
'I know,' I say.
'Est-ce moi?' she whispers.
'Of course,' I lie.

Jeanne will never be as pretty as this doll. Even so, she is beautiful.

David Williams

1756
Llangain, Wales

It's only been forty-five years since Jeanne Breconnier was dying in a hospital in Montreal, her chubby Charles holding her hand and weeping. Her granddaughter Edith (my 6th great grandmother) is ten years old now, working in a Toronto mill and hoping to run away to New York if she can only convince her brother to come with her. It's no good travelling without a man. I have 255 other 6th great grandparents, or had or will have—for some of them are long dead and some not yet born. A century can hold up to six generations or just two, depending on the age of the individual when they become a parent. You get the picture. It's blurry, full of generational overlaps. In any case, way too many direct ancestors to get to know properly, and it's easy to forget each existed. That each had, for example, a preference for pork sausages, the colour lavender, dangerous men, the smell of woodsmoke—or a fear of heights, a phobia about bathtubs, an allergy to macadamia nuts. Everyone who ever existed was, in their time, as real as you and me. You might say *duh*, but be honest—do you really believe the lives of others are as real as yours? It's hard enough to believe when you're in a crowded street—all those other people at the centre of their own orbits! Or looking out of an airplane window at a sprawling city. And long-dead lives can seem especially ephemeral. Maybe it's not possible to allow in so much awareness. Brain space is finite, and the bottom line is other people's lives are mostly ignored and soon forgotten. Which is one of the reasons I visit my ancestors—they get no attention otherwise and I feel sorry for them. My visits began as duty calls, but without exception each ancestor has had me riveted me within minutes. No life form is boring when considered properly. Not a single one. (You listening Kevin? William the worm and Polly the prokaryote? You guys are incredible.)

Another thing, in addition to the empathy I began feeling with ancestors like Tony and Margaret and Jeanne—more and more I find I am able to flit into the consciousness of those who are *not* my ancestors. I can't talk to them, but I get a quick glimpse of what their personal rock face looks like. I'm not omniscient—no one is—these are more like brief flashes of illumination, perhaps bestowed due to their proximity to my ancestor. Along with so much else, no one knows the truth of these things.

~~~

It's 1756 and we're in Llangain, south-west Wales. Ingredients for the Industrial Revolution are simmering on back burners in places like London and Manchester, about to be brought to a boil—but no one here suspects, for hasn't rural life been the same since time immemorial? The king of Great Britain is George II. He prefers long curly grey wigs, tights and garters. He's not terribly popular due to excessive philandering, temper tantrums, and the unforgivable fact he was born and brought up in Germany. America is currently a cluster of European colonies, pro-equality yet oddly anti-Catholic, antisemitic and very pro-slavery. Baroque music is starting to wane in popularity, not that folk are familiar with it here—but everyone knows the words to Sosban Fach, or Little Saucepan. The Methodist revival movement is suppressing Welsh folk music and promoting hymns in English, but some songs just can't be killed.

And of course, humming quietly in the background, Homo sapiens are still evolving because that is the nature of Life. It's a live performance, unique each day. The current stable climate is allowing our body size to slowly increase. We're getting bigger and smarter, and being bigger and smarter gives us a better chance of overcoming complex challenges. Unstable climate, of course, can have the opposite effect. Shrinking bodies and brains. Not good news for our descendants, but let's not think about that now.

All you need to know about Llangain is it's entirely devoid of prosperity but pretty in an undramatic way. Gentle green hills roll down to the River Towy, which soon widens into the Irish Sea. Ships anchor in the estuary to load wool, produce and fish, and unload the things that cannot be produced here. The nearest big town is Carmarthen, considered the oldest town in Wales, and the people in it brag it's the birthplace of King Arthur and Merlin—but so do the people in a dozen other Welsh towns. Most folk in Llangain don't go into Carmarthen, or maybe once in their short lifespan for a festival or market day or a hanging. They're too busy trying to stay above ground.

Look around.

Scattered farms in the crevices between hills, with bedraggled sheep and dark, thatched houses that squat down and blend in with the landscape. Now and then, a whitewashed two-storey building like a child's drawing of a house, very symmetrical and friendly, roses on a trellis, sheepdog and cat in front. But up close, trust me, these houses are full of shadows too. It's early morning and we're on the precipice of winter. Black smoke rises out of chimneys and curls out over fields. Don't breathe too deeply and wrap up warmly. A place like this can suck the joy out of a person in two seconds flat.

We're visiting one of the hunkered-down houses and inside candles are lit because the sun will not hit till nine a.m., and only then if those clouds clear. My 5th great grandfather is

David McWilliam Williams—in this family, you can never have too many Williams. His father's name is William Williams. In twenty-three years, David will name his fifth child (yes, his fifth) William and not think that odd at all. In fact, his first impulse will be to give him the middle name McWilliam, making him William McWilliam Williams, but his young wife Margaret (yes, yet another Margaret) will put an end to that discussion with a snort of derision. And by then, he'll have learned not to bother arguing with her. But David is only one today. He doesn't know any of this. Names are just repeated sounds. If he knew what lies ahead, he'd not believe it. He'd probably hide under his bed for the rest of his life. Or maybe not.

There he is.

'Happy birthday, David,' I say in an extra cheery voice, the one I use when I'm in a cheerless situation.

'S'not my birthday.'

He's a runny-nosed rat-nest-haired skinny thing, sitting in a corner of the room that serves as kitchen, living room and master bedroom. He's very still and quiet for a baby, as if trying to be invisible. Could be a survival tactic. More likely the result of not getting enough attention, so why waste energy trying. He smells sour, and there's not much pleasure in looking at him. I don't often think that of any child, but there you go. I have no urge to gobble him up or even kiss him. He reeks of unlove. In general a child like this does not attract the one thing it needs more than food—affection. I can't stop an image of runts being pushed out of the family nest to die in the cold. Nature's efficient way of giving available resources to the offspring most likely to thrive. His mother Mary-Gwyn loves him, but her love is spread thinly, and David is a child who needs five times that amount. I give my heart a little shake, try to coax some fondness into it…though it does not come naturally. This boy looks grey in every way.

'Actually, it *is* your birthday David. Your first birthday.'

'Humph. My name's not just David, anyway.'

Here we go, I think. Another argumentative ancestor, just what I need. Well, two can play at that game.

'You're right! Silly me. You're, uh, Saint David. Patron saint of Wales, right?'

Quite a long pause. Is he weighing me up?

'Correct,' he finally says, with a trace of a smile. Goofy or ironic? I can't tell.

I smile too, because he's funny to me now. No idea why.

He indicates that I can bow, which I do, because he's Saint David. When I look up again, he's suddenly handsome. Nothing about his appearance has changed, but there is something appealing now in his demeanour. It's uncanny. I have bestowed attention on him, and voila! A hard little bud has bloomed.

'Pardon my ignorance, your Holy Reverence.'

'You may kiss my hand.'

He extends a filthy paw and I kiss it with pleasure as if I really am a grateful and temporarily favoured servant. One of my character flaws is an over-susceptibility to people's behaviour—not just mimicking their personalities, but becoming who they think I am. Some days I feel I have nothing solid at my core. A quivering mass of jelly ready to adapt to whoever I find in front of me. And yes this has got me into a few messy situations. Once I wore a powder blue pantsuit I hated for months because a boy I liked bought it for me. I'm not proud, but then someone started off this genetic chameleon stuff and it wasn't me. Biological imperative of a porous personality? Obviously there are advantages to blending in, of being invisible to predators (ask Harriet the Forgettable), and also mirroring powerful beings in order to flatter them. You're less likely to be killed if you resemble your enemy, or at least intuit what they're thinking. But the disadvantages of a chronic identity crisis are too humiliating and numerous to list here.

'Your Saintliness, please accept my humble wishes for a happy birthday.'

He stares at me, so I repeat myself. 'Penblwydd hapus! Happy birthday.'

'So what, and happy birthday to you too!' he says, as if the phrase is an insult and he's hurling it back. *What you say is what you are, ha ha ha.*

'I'm sorry I've upset you, David.'

'Oh, that's alright.' He's deflated now, back to being uncute—but somehow I'm not repulsed this time. Maybe I can just feel sorry for him now.

'And I'm not really a saint,' he continues. 'Call me David. That's what I'm called. When I'm called anything. Which is almost never.'

'Okay, David.'

I'd be inclined to report this neglect to a social work department, but there's no such thing here. The Poor Law Act gives church wardens responsibility for collecting tithes and distributing them to the poor, but otherwise there's no official protection for the vulnerable.

'You're just saying that because you feel sorry for me.'

Oh dear.

~~~

When David is almost four years old, his wife is born in Ireland. Dungannon, Ulster to be precise—314 miles north-west from Llangain, and forty miles inland from Belfast if you skirt the southern edge of Loch Neagh, which is so big you can't see one side from the other. It's like travelling a seaside coast, and not many Dungannions bother to make the journey because their hometown has everything they need. But let me tell you right now, even so, this is a bleak place. Like Llangain in

Wales, this is not a place you'd choose to be born in. The Ulster Plantation—Catholics in the north of Ireland being forced to surrender their land to British Protestants—ended about a century ago, but it's still a tricky time for Catholics. They will not be allowed to own land, or a horse worth more than £5, or join the army, or become a doctor or lawyer, or worship freely at Mass, or open a Catholic school until 1778 when the Penal Acts are repealed—although even then, the privileges will only be granted once the Catholic individual has signed an oath promising allegiance to the Crown, not the Pope. Irish Catholics will not be able to vote until 1793, and their legal protection against discrimination will not be set in law until 1973. In short, it's not good news for Catholics anywhere in Ireland for several centuries. To their credit, and the frustration of a sequence of British kings and queens, the majority of Irish remain Catholic during the entire period.

~~~

My 5th great grandmother Margaret Lyons is born with a shock of red hair courtesy of Florence from Doggerland, plus seventeen other red haired ancestors including Ned the Neanderthal. (Yes, some Neanderthals definitely had red hair.) She belongs to a Catholic family—which means a clandestine culture of caution. Nevertheless Margaret's mother Bridie keeps the faith, and also keeps their hovel as tidy as she can, given they share the space with sixteen family members, the cow in bad weather, and the hens when they find their way in, the bothersome creatures. If only they'd lay more eggs! Margaret's father is a good man who never gives up on his 3.2 acres of bog, although he is not in fact Margaret's true father. Nor, like at least 3% of fathers worldwide and 6% locally, is he aware of this fact. Bridie lost her head briefly over Father Sweeney (the priest with that adorable chin dimple) at the secret local church one day while doing the flowers—but it's a well buried mistake, even from the priest. He fathered another child that month, who will become a close friend of Margaret's. Though they share a chin dimple, not a single person (including them) will ever know they are half-sisters.

In three years Bridie will put away her rosary beads and tell the priest she'll no longer be doing the flowers. That in fact, she'll no longer be taking her children to Mass on Sunday at all, at all, at all. She begins buying meat on Friday mornings from the butcher where everyone can see her—although never cooks it until Saturday. She renames two of her children who are too young to object—Patrick is now Samuel, and Liam is now Jacob—old testament names being the favourites of Protestants. Luckily she doesn't need to alter their surname, for Lyons has evolved from the Normans and has no Catholic connotations. Bridie hasn't lost her Catholicism, of course—she couldn't if she tried, any more than she could change the colour of her eyes. She's merely guiding her family into the safer waters of Protestantism, in hopes of an easier life. Her husband, who's just a churchgoer—not Catholic in his bones—will be fine with this new state of affairs. They'll begin attending services at the Methodist church, and it's not such a tragedy. Not such a betrayal, Bridie tells herself, for isn't everybody worshipping the same God anyway? Reading from the same bible? Yes, yes, it's vile to hear certain phrases freely bandied about in her presence, like *dirty papist* and *fecking Jacobite*. Very stressful having to keep her face expressionless. But a mother is entitled to save her family, that's not a mortal sin last time she checked. And that is that. The Catholic God is forgiving. Young Margaret, only three by this time, will not remember being baptised Catholic or attending Mass. Her bones will still be too young to have any particular religion in them. And this will save her too.

~~~

Back over the Irish Sea to Wales. David is seven years old now. He's a pale undernourished boy in hand-me-downs, kneeling in the kitchen and praying. How did that contrary one-year-old become so docile? So devout? And my goodness, his face is spattered with more freckles than I've ever seen on a face. He must have a triple helping of the gene variant MC1R leading to a melanin excess. A dozen more freckles and his skin would be entirely dark.

'I love you, God,' he's whispering with a trace of a lisp. 'In the name of the Father, the Son and the Holy Spirit. Amen.'

Maybe it's a defence mechanism because he's a weakling. If you can't fight them, turn the other cheek and bow to avoid being killed. Or maybe God exists—because he might, no proof either way. Maybe David has been born knowing God is real, and he's the smartest cookie in this house of atheist Methodists. His voice is so earnest, so innocent, I get the lump in my throat I always get hearing children sing Silent Night. I remember how childhood faith feels—in particular, Catholicism, because I was taught by Dominican nuns and priests. When I can trick it back into being, it offers pleasingly plump moments of peace. Sometimes hours.

I notice no one notices praying David. The family mills around as if he's not planted in the middle of the kitchen, eyes closed, hands clasped, mumbling *I love you God*. Maybe he does this a lot.

'Shush!' I tell them, but of course no one hears me.

David sways slightly on his knees, and his feverish whispers continue, but softly. I have to kneel next to him to catch what he's saying. He seems to be saying the same prayer over and over, with many amens and *I love you Gods*. Then he goes off into something new.

'Please accept my sacrifice, oh Lord,' he lisps. 'May Mary and all the saints and the martyrs and the angels and archangels be a witness to my promise. From now on I will work only for you, God. Take me to be your instrument for good in the world. Your willing servant, me.' Then he adds, in case God needs help: 'David McWilliams Williams of Llangain. That's near the Towy. Second house on left, up the brae. Amen.'

Jesus. He's only seven! And yet, there is something so sweet about his plea, I want to light a candle and be a better person too. But wait. Now he's pausing, eyes open and tilting his head as if listening to a reply. Or waiting for a reply. His eyes close again, as if savouring an answer. Or as if all the petty annoyances of being seven have fallen away. His frustration with his big brother's habit of hogging the whole bed. His anger at his father for slapping his mother last night. His disappointed vanity at being shorter than his younger brother. Gone, gone, gone into the indifferent ether. *God, come get me! I'm yours!* Or maybe I'm wrong and he's not filled with any kind of certainty or serenity. Maybe he's drowning in a sea of insecurity, and prayer is reaching for a lifeline even when you can't see it—you just keep hoping someone has thrown it.

'Amen,' he repeats, clearing his throat and opening his eyes to look upward sternly.

For his sake, I will any kind of response. A burst of bird song. A shaft of sun light.

I tell the God I miss believing in: If you want to meet a fan, here's your chance.

David coughs and clears his throat a few more times. Ahem!

'Amen, amen, amen,' he chants loudly, almost angrily. 'I said, *Amen!*'

'Will you give it a bloody rest,' one of his many brothers says from the table. Another brother throws a stale loaf end at him.

'Come along David! There'll be no food left, if you don't hurry up,' says Mary-Gwyn, not even bothering to look at him. Her arms are bare and her hands are kneading what looks like grey playdough. Bread, I guess. No wonder everyone around here is so thin.

~~~

Two and a half years later, David is at the door of the very Reverend William Williams, also known simply as Pantycelyn because that's where he lives and he's easily the most famous person there. Not only is the reverend famous for composing hymns and writing poetry, he's a preeminent figure in the Welsh Methodist revival movement. An A list star in this neck of the woods. He lives thirty-two miles north of young David, in the parish of Llanfair-ar-y-bryn. At this time and place, influenced by John Wesley's writings, Methodists are strongly urged towards sobriety, patriarchy, obedience and—for the ones headed to reverend-hood, celibacy. Reverends are allowed to marry, but robustly discouraged from it. And it is this idea of celibacy, of purity and solitude, that David is most infatuated with. It seems to offer a life entirely other than the one he's currently living—a house so peopled, there's little time for spirituality or thinking calmly. He left home early three mornings ago, while it was still dark. He left a note for his mother. Yes, all the Williams children over six can read and write English and Welsh. There's no school here, but Reverend Griffith Jones stayed in the Williams house for almost three months—quite long enough to instil literacy. Teaching children how to read is his life's vocation. By the time of his death, Reverend Jones will have taught roughly half the rural population.

Dear Mother. Do not Worry. I am Fine.

I'm going to visit Reverend William Williams.

Sorry I took four slices of Ham and three Apples and Tommy's jacket and some Money.

Byddaf yn ôl yn fuan!

Affectionately, your son David.

He slept in a cousin's house last night, and now it's midday and he's standing fearless on the doorstep of the manse. He's lost the vestige of cuteness that visited briefly around his eighth birthday, and is back to being awkward-looking. His nose is too big for his face, his freckles over-abundant, and he's still a runt—could pass for six-year-old. He has a new cold sore on his lower lip and a stye on his left eye. His posture suggests humility, but oh my, the determination burning in his soul. Almost an affliction, in these times and parts. Aside, hopefully, from somewhere like here. The Reverend William Williams house.

Knock, knock, knock, go David's tiny brave fists. He's hungry and tired. *Knock, knock knock!* The door, twice his height, finally opens with a slow creak.

'Good morning. And who might you be and where have you come from?' asks William Williams. At sixty-six, he's handsome in a wrinkly neat-featured clean-shaven way, and he's wearing a long black cloak as if on his way out. Maybe he's just finished a bowl of tomato soup, for there's a faint orange tinge around his mouth.

'David McWilliams Williams from Llangain, your eminence.' He whispers this, having lost his confidence. The man is so tall! So imposing! He has a faint hope their similar names will be commented on, break the ice. This does not occur.

'Llangain? But that's quite a long walk, David. What do you want?'

'I'm here to work for you. To live here as your servant.' Because he's rehearsed this speech so often, it comes out with a false casualness. As if his room is waiting, with age nine black garb already lying across a hard-as-stone bed. Then he mumbles: 'I want to give my life to Jesus.' *Creithio fe*, he thinks. It sounded so much better in his head.

'I see. A noble calling. You won't want a wife one day? Children of your own?'

'No!'

'But why?'

'God called me. I only want to serve Him.'

Short pause while the reverend Williams peers up and down the lane to see if any parent is loitering.

'How do you know God called you?' He bends down to ask this, thinking how on earth will he get this little kid back to Llangain. David swallows loudly and the reverend repeats, more gently: 'How do you know, my child?'

'I heard him,' he whispers, eyes downcast.

'Is that the truth?'

Another long pause and the man does not stand up. Crouches closer to David.

'I heard him in here,' says David, thumping his bony chest, and the man straightens up with a smile.

'How old are you?'

'Ten. Almost ten.'

'It's tempting to say yes. I can see you're a good boy and I'm sure God loves you. But you're too young.' This is a lie. Some individuals have been in the service of the church since the age of seven.

'But, God said...'

'Tell him your mother needs you a bit longer.'

'But...'

'And you never know about the wife. You might meet a girl one day and change your mind about celibacy.'

'But I don't want...'

'I'll walk you part of the way home, my son. Wait till I change my shoes.'

~~~

That's how close I come to non-existence again. What are the odds I'll exist one day? According to Dr Ali Binazir, a man from Los Angeles with the bizarre title of Happiness Engineer, our chances of existence today are one in 10 followed by 2.6 million zeros. According to the only slightly more credible website called Science Alert, the chances of you being you are one in 400,000,000,000,000,000. There are some who claim the chances of your existence are so infinitesimal as to be as good as zero. Only a fool would bet on my appearance in 1953 as a result of beings not dying or joining celibate sects before they reproduce. If Deacon Reid had answered the door instead of Reverend Williams, then David might have been granted a religious internship in return for a lifetime of unpaid servitude. To be fair, he might also have lived far longer and experienced more serenity than if he'd not joined. There's something to be said for avoiding sex, parenthood, drugs, alcohol, cigarettes and other vices. In 2003, David Snowdon found that nuns were 27% more likely to live into their seventies than the general population.

~~~

The next year, just before David's eleventh birthday, everything changes. David's family, along with three other local families, is scooped up and sent to the north of Ireland. The scooper-upper is the Welsh landowner to whom they pay a tithe annually. He's bought an Irish estate and prefers his properties to be occupied by fellow Welshmen. The three families travel with their most treasured belongings via horse-drawn carts, a ship, and carts again on the other side of the Irish Sea. The journey takes over a month, and no one expects the destination to be an improvement over the life they've been yanked out of. The general mood is one of stoicism, of endured discomfort and ebbing hope. But on arrival in County Tyrone, David's father is assigned a bigger farm (livestock and tools included) than his Welsh property. It's in a valley not three miles from Dungannon, where David's future wife—now aged seven—lives. Despite their proximity, they will not meet properly for years, though they'll pass within inches of each other on market days, and Margaret will develop a little crush on David which will simmer until they have an occasion to speak three years hence. Yes, seven-year-old girls can have crushes—a platonic romantic yearning or even obsession. Children often fall in love with a parent, for instance, perhaps as a trial run for later sexual relationships. But Margaret's crush is not because David looks like he'll grow into a tall handsome man—that will never happen—but because she spots him crossing himself when he sees an old woman's body being carried away from a shabby house in the town. Doing the sign of the cross is not a Protestant act generally, but the Methodists do it. Perhaps for Margaret, the sight has a comforting Catholic connotation, although consciously she has no memory of being Catholic. Then he spontaneously offers his arm to the old man following the corpse with a lost look. Doesn't hesitate, even though he's no doubt due to help his father sell hogs at the auction which begins in five minutes. Such unexpected kindness from a rough farm boy! She also likes him because there's something about his freckles, his serious pale eyes and non-swaggering walk. Margaret's father is not a swaggering man either—he relies on his wife Bridie unequivocally and shamelessly—but her brothers are, and she senses swagger boys are to be avoided. Although all her friends giggle when encountering one, as if cocky boys are the bee's knees. Those boys scare my 5th great grandmother.

~~~

Ireland seems like a dream for the Williams family, for the property includes a fabulous two-storey three-bedroomed house. It still has furniture and pots and pans, even some pairs of boots, all of which presumably belonged to the Catholic O'Hara family who are now homeless in the more southern counties, exiled for their faith. It's not a situation David overly judges. He experiences an uneasiness at first, wearing a stranger's boots, but doesn't attribute this to guilt. It's hard to think clearly about the rights and wrongs of a thing when everyone is doing it and you benefit. After a relatively short while, the Williams family settle in and life approaches normality. Mary-Gwyn and her husband make new friends, find the closest Methodist church, plant their first potato crop, butcher their first hogs, figure out some Irish Gaelic words. Not easy, for though Welsh and Irish are both Celtic languages, they've little in common.

~~~

It looks like this might be the end of the story for the Williams family—or the beginning of the end, for if you know where you are going to get old and die, you can shut down and cruise. Surrender to routines, the expected milestones and petty detours, and even relish them. But fast forward several years to a May afternoon, and Mary-Gwyn is bustling home from the market with the news that a ship bound for Philadelphia is offering berths to families for free. Apparently, she explains to her husband, American Presbyterians have forged a link with the Presbyterian church here in Ulster. Not only that, they're offering free frontier land.

'Enormous amounts of land for hard-working families like us, William. That's what the mannie was telling us all today,' she says. Her face has a look he recognises and somewhat fears. A feverish excitement, an intentness, and he knows she will not shut up about this until she's had her way.

'But we're not Presbyterians. And why would you want to go to the colonies? They're so far away, we'd know no one there. We've a great life here, Mary-Gwyn.' He's just come in too, from the barn, and is dying to tell her about the calving.

How great it's going. Three heifer calves and a bull. By God, it's finally looking good for the cattle sales next season, and wasn't he clever to get rid of the hogs?

'Are you mad?' she asks, exasperated. 'Look around!'

He looks around dutifully but everything he sees pleases him. The lit range, the shirts hanging to dry above it, the three new loaves rising with a cloth covering them.

'I love this place,' he whispers, overcome. 'I do, Mary-Gwyn.'

'Ah, but it's not ours is it? It'll never belong to us, William. We didn't choose it, and it could be taken away from us anytime with no warning. We'd be fools not to grab a chance of owning our own land. Think of it! And it doesn't matter a whit that we're not Presbyterians, as long as we're some kind of Protestant.'

Yes, there is more than a little nomadic DNA from Freddie the Kenyan hominid at work in Mary-Gwyn's blood. And risk-taking DNA from Margaret the fish who decided to try life on land. And just plain territorialism, present in all Homo sapiens but especially abundant in Mary-Gwyn. These genes run strong in her family, not always to a sound purpose. Her poor husband slides his bonnet off and sits down. Sighs. He only has so many more big changes in him. Such a vast measure of energy they each take. What kind of weather would they even have over there?

'Think of the children!' says his wife.

But all he can think of is shallow graves in foreign soil. Of having to learn new names for things and where things are. Of packing up again.

~~~

Not three miles away, a similar scene is being enacted in Margaret Lyon's kitchen. She's feeding the fire while her mother Bridie pummels her father with arguments for emigrating. Margaret listens nervously. The air feels thick, and it's not easy to take a breath. Finally her mother stops talking. Sighs. Sighs again. When her father remains silent, her mother opens her mouth and out comes the big guns. The pronouncement on personal destiny she often saves for last. She employs the tone she always uses at this stage. Low, mean, almost a hiss.

'Listen here, you pathetic piece of lard. Our children are destined for greatness, they are different, mark my words. And we ARE NOT, I repeat NOT, going to let this opportunity slip away from us.'

Margaret feels a thin thrill sliver its way into her stomach, chest, right out to all her extremities until her whole being tingles. Destined for greatness? Yes, that will suit her very well, thanks. She grows a tenth of an inch taller simply with the knowledge of her innate superiority.

~~~

Next week fourteen-year-old David—still scrawny and short, still over-freckled but washed up nicely—boards a sailing ship in Londonderry bound for Philadelphia with his family. His family being his mother Mary-Gwyn, his father, four brothers, six sisters, and a recently orphaned infant cousin from his mother's side called Rebecca. Mary-Gwyn marches her family up the gangplank, head high, broad shoulders squared, Rebecca strapped howling to her chest. William takes up the rear, head down, back bowed under a large rectangular case tied with belts. Each Williams carries a bag or two with their belongings inside—knives, hairbrushes, bloomers and socks jostling with mementos and heirlooms like a maiden aunt's necklace, a vial of holy water, a lace shawl. They are shown their quarters—not a private room as Mary-Gwyn had intimated, but narrow hammocks in rows, hung between decks with no windows and the ceiling not high enough for anyone to stand, aside from the three youngest. Their hammocks are surrounded by over 200 others. It's a dark vast dormitory.

'Where's the baby going to sleep?' David asks his mother, counting beds.

'With me, with you, with anyone. What's it matter? She takes up no room.' A transparent lie, for not only is Rebecca large for her age, she's also a terrible sleeper.

'Iawn,' says David quietly in Welsh, trying not to notice the look of uncertainty in his mother's eyes. *Alright.* He began praying this morning before he got out of bed, and has not stopped since. Even while talking or listening, the praying goes on. So much praying causes a mental groove to form, along which the prayer circulates with minimum impetus needed. He's not always conscious of the words, and it's possible the words have ceased to have meaning. Oft repeated words often do. Nevertheless, the meaning under the words is loud and clear today: *Help us. Help me.*

Meanwhile, mere yards away, ten-year-old Margaret Lyons is trying out her new bed. It's windy even inside the Londonderry harbour and she's not happy about the rocking motion, although luckily her stomach is not affected. Bridie doesn't notice, for she's too caught up in the moment. The Lyons are on their way to their brand-new life, hip hip hooray! God save them (as well as forgive them for jumping the Papal ship), for each one of them is destined for greatness. Oh, the world's never been scarier or more exciting. Margaret's friend Margaret Leary, is in the hammock next to her, her mother having died three months ago giving birth. Her father disappeared, last seen in a pub, and the Leary children have been distributed here and there and everywhere, with Bridie taking in young Margaret. What's one more? She'll keep an eye on her until she gets the girl settled somewhere in the colonies. A post as a housemaid in Philadelphia, hopefully. Ten is young, but not too young. The current child labour law in Ireland specifies apprentices cannot be younger than eight years old, and schooling is mandatory for no one. Ten is quite old enough to earn a living.

Not everyone around here feels obliged to take in orphans and waifs. Bridie and Mary-Gwyn are hospitable to a fault. They'd probably open their door to ghosts, if they came knocking. They'd fling their arms wide open and say: *Aw, poor you, rotten luck to be dead. Come on in! Plenty room, make yourself at home.*

~~~

The two Margarets share a chin dimple and good singing voices, courtesy of flirty Father Sweeny, but little else. Margaret Lyons is tiny for her age, with red hair so curly it seems aggressive. Margaret Leary, on the other hand, is a good six inches taller and more robust in every way. Her hair is a no-nonsense brown and straight. They both wear their hair in plaits, and from the moment of embarkation, they're inseparable, often giggling.

Are they afraid? Is David afraid? You bet. Everyone on this ship is at least a little afraid, even the crew. Secretly, even Captain Mack, a sturdy bearded man of fifty who's sailed for three decades. It's a well-known fact that ships never arrive with the same number of passengers they set sail with. Terrible things can happen at sea. Hurricanes, disease, pirates (yes, pirates!), food and fresh water scarcity, doldrums that leave the ship's sails empty for weeks, and occasionally food poisoning. The trip takes at least seven weeks, sometimes up to three months, and it's likely 20% of the passengers currently checking their new beds for bugs will never arrive. But most passengers have nothing to lose, and anyway disease is rife at home too. It's an exciting day and though worried, momentarily no one believes tragedy will happen to them. So off they all sail into the wide unknown. The *Killarney Rose* is a beautiful wooden schooner with thirteen sails, all filling nicely with an easterly wind, and the tide is on their side. Bon voyage young Margaret Lyon, sucking the end of your red plait! Be brave, motherless Margaret Leary! Good luck full-of-secret-prayers David!

~~~

David and the Margarets quickly become close friends, in the way people do when they are stuck somewhere together. Margaret Lyons is a short-ass tomboy, but David thinks she'll be pretty soon, maybe when she's thirteen. She reminds him of a mongrel dog he had once—cheeky, annoying, fun. Her forthrightness makes him laugh, and she has a way of singing softly to herself that touches him. Margaret Leary towers over him, which inclines him to think of her as older—but as soon as she opens her mouth, her childishness dispels this idea. She's sweeter than Margaret Lyon, softer, shyer, and he likes the way she listens to him seriously and doesn't tease. My Margaret remembers the time she spotted him making the sign of the cross upon seeing the old woman dead, and then offering his arm to the widower. There is kindness in David, and my Margaret is a sucker for kindness. Both girls, in their different ways, are besotted with David.

~~~

Steerage passengers are given ingredients to cook for themselves, and the women barter for prime times at the huge coal-burning range. Mary-Gwyn and Bridie have completely different ideas about cooking. Bridie boils everything until it's a second from dissolving. Mary-Gwyn likes to coat things in salted flour and fry them. Water is rationed, and what isn't drunk is used multiple times before disposal—washing bodies, washing clothes, soaking beans, cooking, washing dishes. Its final destination is usually the bucket used for a toilet behind a curtain, before the contents are swilled overboard. Some days David and the Margarets fight over the right to empty the bucket, for it means time above on deck within sight of the first-class passengers and the crew and all that light and fresh air.

Babies are born of course—it's inevitable, on an emigrant ship full of families. About a quarter die at birth or soon after. Cholera and tuberculosis have been circulating, taking their toll, which brings us to funerals—also an inevitable event on board. Sometimes the sea burials are unwitnessed, and it's only when a hammock becomes vacant that the death is noted. In such cases it's likely that the individual has not been claimed by anyone, or isn't deemed important enough, or the weather is horrendous. Other times there are services, short sombre affairs. Often everyone on deck together, steerage and private. Captain Mack, or sometimes the first officer, says a few words about the deceased—name, birth date and place of origin. Occupation, claim to fame if any. Name of spouse and number of children, if these exist. Then he reads from the Bible—*Even though I walk through the valley of the shadow of death, I will fear no evil*. Heads bow, sometimes tears slide off faces and join the salty air. If the weather is fair and the fiddlers (the ship is stuffed with fiddlers) are in the mood, they accompany the singing of a hymn, followed by The Parting Glass. The Margarets know they can sing, and see no need for modesty. They always belt it out, their voices clear and wavery at the right spots, weaving into the fiddle strains. By the end of the song they're often the only ones singing, for everyone wants to hear their voices. Even the fiddles grow soft. One more prayer, then the dead person, probably wrapped in the blanket they brought from home, is dropped over the stern like a sack of potatoes—and that is that. Back to business.

~~~

Time passes, even while it seems not to. There's a monotony

at sea, and weeks feel like years. Faces and places left behind grow dim, fade into unreality. Everybody's hair grows wild, oily, knotted. Beards appear, skin burns and roughens, lips chap. By the time Pennsylvania is in sight (*Land ho!* shouts a twelve-year-old named Jeremiah perched in the crow's nest lookout), the ship is considerably lighter in terms of human freight. Unfortunately it's almost dark and they've had the rotten luck to arrive at the same time as a hurricane. Well, it's hurricane season, so not too surprising—they've been lucky so far on this trip. Hurricanes will be officially nameless for another two hundred years—nevertheless, sailors tend to consider them female and give them personalities.

'God's blood, but she's a right old blowsabella!' says one old salt named Henry, trying to grab a line that keeps flicking and wriggling like a snake.

'Ay, she's a bobtail alright,' agrees his mate, but without feeling. Hey ho, another hurricane.

And here they are, in sight of America, crazy wind or not. Hooray! Everyone has packed up their bags and washed themselves as well as they're able, and in David's case, said extra prayers for a safe landing. The emigrants stand huddled against the gale on the top deck, each crammed with specific visions of the future. One man has already visualised the hogs he's going to raise, because he's brought the breeding pair with him and the female is already pregnant. He'll build a shed for slaughtering and butchering, a shed for smoking, but he'll not need a pen—they'll roam freely in his own woods, and the money he'll get at market for bacon and hams will keep him happy for the rest of his life. In unison the spray-soaked emigrants tilt against the swells, compensating to stay upright. Captain Mack, also tilting, stands in front of them frowning. He shouts:

'We'll not get ashore tonight. Not in this wind, and too late in the day. We'll sail up the Delaware in the morning. Go back to your cabins! Go back below!'

Is he slurring? He sounds odd. They grumble and want to stay on deck anyway, peer at the dark line of land. Some defy the orders—but David goes below because his mother has gone below, and the Margarets follow him because that is the dynamic of their friendship.

At least seven events (five meteorological, two nautical) contribute to the disaster about to happen—but the most fatal, the most preventable, is the amount of rum in Captain Mack's stomach. He is drunk. He's so drunk, he stumbles twice on his way back to the bridge. When he gets there, he grips the helm so tightly his hands turn white. He's had a toothache for three days. He has not slept or eaten, and he cannot think aside from hatching the brilliant idea of asking his first mate to cut off his head. Remove it and toss it overboard! He shouts the order to take all the sails down bar two. (He should've shouted this an hour ago.) Sailors scamper up the three wildly rocking masts to begin unfurling eleven sails. But the hurricane, the blowsabella (which like the emigrants has travelled far, coming from the Cape Verdean islands off West Africa) means wind is gusting, not just blowing. One of these gusts punches the biggest sail just before it comes down and just as the ship has dived into the deep trough of a wave. Water pours over the port-side gunnel and down into the hold. Two seconds later, the deepest part of the hull has confused fish swimming around in it. The ship has no place to go but abruptly sideways. Can this be reversed? Rectified? It cannot, and the ship is unequivocally capsized. Captain Mack curses and berates himself in fury as the floor becomes a wall and the wall becomes a floor. When he stops sliding, his face is pressed into a framed picture of King George. Captain Mack's nose is broken and it's such a distracting sensation, his tooth ceases aching momentarily. The sailors on the booms are now in the sea. Everyone and every loose thing on the top deck is also in the water. Bridie with her hidden rosary beads, and her husband (who dies nameless in this narrative, I'm sorry to say), and all their children minus their Margaret and Margaret Leary. The fiddlers, other children and babies, other wives and husbands. Young single men bound for the frontier and young single women bound for marriage or service. Emigrants of all shapes and sizes—into the sea they all go. The twenty-two crew. The cooks. Captain Mack, who has finally crawled out of the bridge, dazed and red-nosed, and slid straight into the brine. The first mate. The second mate. The ship's three cats, yowling. Luggage, boxes, bags, hens, barrels, hats, eyeglasses, handkerchiefs, wedding rings, two hogs, barrels of Irish whiskey, pipes, pouches, gold coins, lace shawls, three bags of apple seeds, twenty-eight gold pocket watches, 298 miniature paintings of loved ones, bars of lye soap and string-tied letters. Down, down into the dark waves. Down go the imagined futures—totted up savings and likely expenses, houses successfully built, curtains sewn, babies born, triumphant return visits home to Ireland. Most folk disappear instantly. A few bob and flail, pale smears on the water. Below deck, in the sideways dark steerage deck, everyone is screaming but baby Rebecca who was thrown so hard her head collided with a metal pipe causing instant death. Mary-Gwyn is frowning, looks startled, as if just realising a massive miscalculation—but no one sees her face because it's pitch black. Everyone screams or prays or cries or dies.

    David feels his way up the wall which used to be floor, finds the hatch and something to brace his feet on, then he pushes the hatch open. The deck is now vertical and nobody is in sight. He peers down into the churning sea straight below him and clings fast to the hatch frame. The sails are slapping the waves and the halyards are snapping against the horizontal masts, but there is a curious absence of human noise up here. Everyone is in the sea, and the noise coming out of their panicked mouths is lost in the rush of wind and sea. *Huh*, thinks a stunned David. *And I've had no chance to be a martyr yet, or a reverend or even a deacon.* Sainthood seems unavailable now too.

'Our Father who art in heaven, hallowed be thy name,' he shouts into the wind, as if words can deliver him. Form a rescuing line to swing him off this ship and into heaven. Should he let go, allow himself to slip into the water and be done with it? He feels a warm calm enter his chest, a sense of rightness. A lightness in his limbs. Of course he's going to die today. *On my way, God! Get the kettle on.* (He has no idea the consequences of his death, of course. Not a clue he's about to axe the possibility of me. Not saying that'd be a huge loss to the world, just saying I'm very close to non-existence again.)

'Get out of the way,' screams Margaret Lyons, shoving her way up below him like a trapped cat discovering an escape route. After a second, he makes room for her in the hatch opening, and then offers a hand to her friend, the other Margaret. It's a very tight squeeze now, and a ludicrous sight—the three heads and torsos sticking sideways out of the ship's deck like a three headed creature. Roiling sea below them, quarter-moon-lit sky above, the air wild with wind and salt spray. Because they're young, they can't help feeling a sliver of exhilaration thread through their dread, as if their brains haven't entirely understood the seriousness. *Blood and hounds, but this is exciting!* Below deck the screams begin to quieten, turn into voices plaintively calling names of loved ones, and sometimes an answer comes: *I'm over here! I can't move!* Much moaning and weeping. The three children are stoppers at the bottleneck of hell. David recommences praying while the girls in unison burst into song.

'Guide me now oh Lord, at the hour of need, guide me to your house and...' calls out David.

'Be thou my vision, oh Lord of my heart...' sing the girls, chins lifted, long hair falling sideways.

Their voices as a bulwark are flagrantly ineffective, so they cease after forty seconds.

Then Margaret Lyons shouts:

'We've got to climb up the deck. I mind there's a boat hanging off the other side.'

'No!' shouts David. 'Too risky.'

'Suit yourself. Move over.'

'There's nothing to hold on to. You'll just slip off the…'

But she's already gone, pulling herself along a line he'd not noticed. She's scrambling up like a monkey, hand over hand, legs wrapped round the line. So he helps the other Margaret do that too. Holds on to her legs until she's got hold of the line. Up she goes, until the two girls straddle the deck rail above him, their hair wild and their skirts and petticoats flapping. He's never been a physically brave boy, but he forces himself from his perch. One last plea to God, and he's out. Once he has the line in his hands, it's not easy for him to pull himself up. But pull he does, and soon the three of them are atop the rail. On one side, the deck going straight down into the sea, and on the other, the dark underside curve of hull and…miracle! There really is a small boat dangling just below them, banging against the hull with every wave. There must be cleats somewhere with the lines attaching the boat to the ship, but he can't see them. It's too dark and everything is slippery.

'Now what?' he cries, hearing the whine in his voice and feeling shame for it.

'We cut it!' cries Margaret Lyons, pulling a knife from her pocket.

Both girls give him mad grins, they even squeal a little. Margaret Lyons begins sawing at the lines, while the other Margaret hooks her fingers to her friend's corset. (Yes, children are wearing corsets already—not for fashion, but to strengthen their back for manual labour.) David begins to see the problem before they do, for when the lines are severed, the boat flies into the sea without them. Off it bobs, upside down, and soon disappears in the black swells. And all along, the roar of the wind, the churning enormous waves which break on the hull and otherwise remain intact like moving mountain ranges. The surprisingly loud creaks and groans of the ship's timbers, as if it's a hurt animal. The cries from below deck, of humans who understand they've arrived at their last moments. Minutes ago David was making the same kind surrendering sob—he was even looking forward to meeting God in person, shaking his hand. But he's changed heart, infected with the girls' determination. They will not die. Absolutely not! Their age and imagination forbid it. Of course, optimism cannot prevent catastrophe, it just maximises the time a life form remains calm—but this calm can aid self-preservation. One biological imperative of optimism—a creature is more likely to figure out a solution because they're not panicking. (Just ask Margaret the fish, arriving on land with the calm assumption her vital organs would cope without gills. If she'd been nervous, panting, needing more oxygen, you'd not be here reading this.) Optimism also accounts for 93% of marriages and subsequent conceptions—for if individuals fully understood the vast variety of heartache, bewilderment and impoverishment a family can inflict, they'd never marry. No, no, no! Brides and grooms need to believe love lasts, or they'd never do it. On a personal level, optimism keeps pulling me back from bleakness, and puts a positive spin on people and situations I would otherwise find unbearable. I don't know if optimism is ever fully warranted, but I know this: It feels better than realism. If the end product of a bad situation is the same no matter how it's viewed, why not take the feel-good view—at least until the tragedy evolves. I have many friends who disagree, and my sister disagrees too.

~~~

'Jump! We need to jump off!' David shouts, because that is what God has just told him to do. Yes, God's voice loud and clear in David's ear at last! He has a very slight Dublin accent, possibly privately educated. It's a nice intelligent voice.

'Jump, jump!' screams David maniacally, like a girl.

The Margarets stare at him.

'No! Wait!' he says. 'Take off your shawls first. And boots. They'll weigh you down.'

Off come the shawls and unlaced boots, which skid down the deck and splash into the sea.

'Can you swim?'

The girls both nod yes. Not because it's true, but because they understand learning how to swim is not an option now. They each recollect everything they ever learned about not drowning. Spread limbs out, hold breath, relax. Or is it best to kick and swing arms madly? Damn, damn, damn!

'Alright,' he shouts. 'I'll go first. I'll slide down to the keel and wait for you. Then we've got to swim away from the ship. It'll suck us all down otherwise. It's sinking.'

Now it's his turn to be the ingenious brave one, but with possibly two drowned girls on his conscience. He doesn't let himself think about it. Instead he thinks—*Did God really just talk to me? He did! God talked to me. To ME. At last.* He doesn't know this, but they are now in the eye of the hurricane. The sea is enjoying a breather before the eye passes. All David knows is that the waves are suddenly calmer. He slides down the hull to the keel on a spiritual high. The water is a shock, though it's far warmer than the North Sea or Lough Neagh. The girls slide screaming down and bump him off the keel. There's nothing for them to hold on to but encrusted barnacles. They spread their selves over the keel anyway, hugging it but slipping with each wave that rises. They're crying with full abandon. David spies a wooden pallet that's somehow made it to this side of the ship (*miracle number one!*), and flings himself towards it. Tows it back to the keel and orders the girls to grab it, which they do, because he's channelling God's authoritative voice now. He's swallowed some sea and his eyes are stinging, but he keeps telling himself: *God talked to me, to me, to ME!*

'Now, kick! Kick!' he orders them. They're still sobbing, but they swivel and start kicking. Slowly, slowly, up and over each swell they move away from the ship. It's not looking so dark anymore, their eyes are adjusting, and it's surprising how quickly they've acclimatised to holding their breath when waves wash over them. Suddenly the hull rolls over so the ship is entirely upside down and the dark keel rises like a giant shark's fin. A moment later the ship seems to take a breath, rise up on an inhalation, before silently and undramatically disappearing. Balloop! The sea has swallowed the *Killarney Rose*. Not cruelly—it's just the sea, and indifferent—but fatally.

David and the Margarets are children, although most of the time they wouldn't consider themselves as such—and neither would their parents, if they still had them. But they are children and now they howl like babies. They're floating in a dark stormy sea off a foreign shore and everyone they love is gone. Then the crying evolves into other things.

'Help us oh Lord in our hour of need, please guide our souls to our final resting, and bless us, and...'

'Whawhawhat the feck are you d,d,doing?' shouts my Margaret, her teeth chattering.

'Praying,' says David, annoyed at the interruption. His teeth are not chattering.

'Are you insane? Praying? Le,le,let's find the b,b,boat.'

'Yes. Good plan. Kick!'

After ten minutes, they spot the boat. (Miracle number two!) They flip it over. Clamber in. The eye of the hurricane passes and the wild winds return. They paddle with their hands, then after a while they huddle together in the bowel of the boat, where there's three inches of sea swishing around. At first they try scooping it out with their hands, but soon give up. Eventually the calamity is broken up, as all dramatic events are, into increments of banality. The cycle of anxiety (Will we drown? Freeze to death? Float lost forever and live on raw fish? Close-eyes-hold-breath-here-comes-another-wave. Will we drown? Freeze to death?) becomes habitual—and anyway, soon there's simply no energy for fear or imagination. Without discussing it, they lie down, keeping their faces out of the water, and press together for warmth. Margaret Lyons is in the middle, which makes sense given she's the smallest and needs heat the most—but it's accidental wisdom. Time keeps passing strangely, in elongated slow-motion warps they sink into—but they've ceased to feel alarmed with the extreme peaks and troughs, with the soakings. More time passes. No more words, and they slip into unconsciousness, a sleep very like the starting notes of death. Separately, they occasionally startle back to being alive with big shuddering breaths, then slump into shallow infrequent breaths again. The night ends, the tide delivers the boat to shore. Their hearts beat.

~~~

I realise this storyline would not work as a plot. It's too contrived, too corny. Too sentimental. But this is not fiction, so coincidence and miraculous events cannot be judged in those terms. This is the true story of my direct ancestors. There they are, two of my 5th great grandparents plus robust Margaret Leary, and they are verifiably the only survivors of a shipwreck. But one moment, please, to honour the drowned emigrants. They lost their chance to die in bed as old people in the middle of happy dreams. They were unlucky. What to do with such unfairness? Why, I believe Mary-Gwyn and Bridie (after five minutes of tears, of dramatic wailing) would say we should live as well as we can on their behalf, if not our own. That is the obligation attached to survival. Live well and look after one's descendants. Oh, and anyone else who needs looking after. Whether or not David and the Margarets are aware of this obligation is yet to be seen.

~~~

The ship capsizing has been noted from shore, but no one knows what ship it is. Families in Ireland will believe their emigrant members are off adventuring in the New World for months to come. Some will never learn of the shipwreck, will never stop daydreaming of a reunion. A triumphant son or daughter, well turned out, entering their old homes and distributing gifts from America. *It's grand to be home again, so it is, but you'd never believe how good life is over there.* On their deathbed they'll still think this and be consoled. Truth can be overrated.

~~~

A week later, wearing clothes donated by the wife of the Yorkshire fisherman who found their boat beached at a sandbar near the mouth of the estuary *(Well, I'll be cow-kicked! Are ye from yon wreck?)*, they're standing on a busy street in central Philadelphia. Drum roll! For the New World colonialists, this is the centre of the universe. New York is marginally bigger, but Philadelphia is where it's all happening. Commerce, politics, revolutionary rumblings. It's a very hot day in late July. Several years have passed in the last few days, entire lifetimes since these three children left Ireland with their families and high hopes. They're on the corner of Market Street and North 7th, stunned into silence and reaching for each other's hands—skinny David in the middle, little Margaret leaning into him and statuesque Margaret leaning over him. David's trousers are rolled up at the cuffs and belted with hemp cord. Margaret Lyon's dress is bunched up over an old belt to keep the hem out of the mud, and Margaret Leary's dress is faded and patched but the right length. All three are wearing oversized boots that have seen better days. They look younger than they are, although it's obvious they've left childhood far behind. Look at them. Orphaned waifs adrift in the world without a penny, just a piece of paper with an address scribbled on it. They look down the street of red brick houses and businesses, of well-dressed citizens, and each thinks the same thing: How is this going to work? They've never felt so sealed off from humanity. They start to notice suspicious glances, and so move on. David asks a corner newspaper vendor (*Read all about it in the Gazette! More Redcoats coming!*) for directions to The Friends' Almhouse.

'You're not far. Turn right at the next corner, then keep on till you hit Walnut. Then turn left. The Almhouse is between 3rd and 4th.' He looks worried as he says this. Indecisive. Is it smart to help these kids find a workhouse?

'Thank you.'

'You'll get fed, any road, but it ain't a hostel. Just warning you. Not all them Quakers are soft, you ken?'

Soon they're in a part of the city that's not so prosperous looking. It feels like a place with no grownups in charge. Gone are the straight lines of red brick buildings and clean walkways. It's rough and ready here. Steaming manure. Coarse, make-shift walkways. Half-painted ramshackle buildings. The citizens are a jumble of ethnicities and languages and fashions and skin tones and hair styles and states of health—but overall seem more or less well-nourished. Aside from the Africans, who are hungry-looking and seem to be everywhere trailing respectfully behind masters and mistresses or walking alone head down. Slavery is common in northern Europe, has been common for two centuries—but the children have never seen so many slaves before.

They continue to make their way slowly, sniff the air like wary animals, try to get a sense of the place. Cinnamon, salt, hemp, tar, horse, sweat, sage. Just a wisp of excrement now and then, nowhere near as stinky as Irish towns, although the veneer of civilisation seems far thinner here. They've not witnessed anything untoward yet, but there's an unmistakable air of anarchy and amorality. This is an impression they receive in their guts and store in their cerebral survival arsenal, along with how to make a tourniquet, what to do when you get the runs, and in the Margarets' cases—how to deal with men who try to have their illicit way. For although they are only ten, they understand their value and vulnerability. They have no parents, and their only protector is a crazy boy who thinks it's fine to pray out loud no matter where he is. Who has conversations with God about everything. The weather, the state of his shoes, the chances of not going to bed with an empty stomach tonight. No, the girls need to stay on their toes. Because my goodness, there are a lot of men here! A seething ocean of them. Some all in black with broad brimmed hats pulled down over ponytails. Men in worker's clothes, grubby and grey. Men wearing white stockings under breeches. Waistcoats are very popular, as are shiny jacket buttons and buckles on shoes. And there are some British soldiers here and there—for Parliament has been dispatching regiments since colonial resistance to The New Act (disallowing voting) became organised and violent. Have you guessed that Philadelphia is not a proud outpost of the British Empire? There are rumblings here and whisperings and complicated looks and hard-to-read faces. The soldiers—some looking too young—march smartly in pairs, patrolling. They wear red uniforms, have muskets strapped to their backs, and the studs on the soles of their shiny boots click the cobblestones (when there are some) assertively as if saying *Out of the way! We're still the boss here!* Some soldiers stroll in groups, off duty, bantering about the local women or the new tavern that makes steak pies like ones in London. Some soldiers swagger solo, scowling or singing or swearing, drink in their belly even though it's still morning. Some wander lost, perhaps because the city is still largely unknown to them. Down in Boston, in less than three months, the defiance will become much bolder. British soldiers will open fire and kill five colonialists—which is tragic, but the real significance of the event is that the British soldiers culpable are put on trial for the deaths. Things like this drive King George wild.

Wherever there are soldiers—or any large group of men without women—there are prostitutes. The Margarets are not naïve, but they're shocked at some of the street walkers' lack of inhibition. David blushes and looks away. Lips that red and cheeks that rouged and cleavages that exposed…well!

'Holy mother of God,' whispers Margaret Lyons.

'Stop staring, Magpie!' hisses Margaret.

She's started calling her Magpie because this morning Margaret Lyons became a thief. Already, her pockets are full of other people's bits and pieces. Nothing of value. A bottle top from the street, a half rotten apple from the fisherman's shed, a shiny gold button found in his wife's sewing basket.

'I'm not staring.'

'And for the love of Christ almighty, don't look at the men either.'

'So, and where am I to put my eyes? The sky?'

'The ground, dough-head!'

'This is it. We're here,' says David.

'Looks like a normal house,' says my Margaret.

'Look at the sign. Friends Almhouse and meeting place. Quakers and all denominations including nontheists welcome.'

'What's a nontheist?'

'Someone who fervently believes there's no god. Or gods.'

'Sort of the anti-you, then. Knock on the door, David. I'm starving.'

The Friends Almshouse looks like a respectable place from this angle. The walls are alternating red and yellow bricks, so the overall effect is checkerboard. It's one and a half storeys with a pitched roof, out of which one dormer window pokes. A comfortable bench with curved arms sits under the two large front windows, presumably inviting passers-by to have a little rest. The windows are wide open, presumably to let the July air circulate inside. A large trellis covers one gable end, on which a pear tree has gracefully submitted to be trained. The children have never seen anything like it. It's grand, but it's also weirdly miniature, like a doll's house only bigger.

No one answers David's knock on the open door, so they wander to the side of the house and see two rather large and unpretty buildings situated in a field. One is the workhouse and one is the poorhouse. Men, women and children are wandering around between them, talking in low voices. A woman wearing a drab dress and black bonnet tied under her chin directs them to the poorhouse, where they are given bowls of gristly mutton stew, cups of tea, hunks of bread. This cheers them up, but not much. Since nearly drowning, their emotional range has narrowed. As if survival hasn't sunk in at some level, and they are still conserving every available gram of energy. After the meal, they feel full and dull. Their beds are in the Almshouse until it is decided what to do with them. Place them elsewhere or put them to the workhouse. The poorhouse is full up and they're too old to put out for adoption.

~~~

Pennsylvania is not your typical New World colony. Oh no! About eighty years ago, King Charles decided to give this land to nobleman Lord Penn, as a way to pay a debt. No matter that the land was not his to give, for goodness sake—after all, who were the natives, the Lenapes? Just some non-English speaking savages with dubious dress sense. Still, things improved for a while when Lord Penn gave the running of the colony to his son William Penn—because this boy was a Quaker, which means he was a rebel. He did not believe in war or violence of any kind. He set out to establish a place where everyone was treated equally, regardless of religion or race, and he drew up a treaty with the Lenapes. Sadly, after he died his two sons hired some Iroquois, and that was the end of the Lenapes in Pennsylvania. Even so, this colony has a reputation for being liberal—and liberal is not a compliment in the eyes of many European settlers.

David is keen to attend a Quaker meeting, for since God talked to him personally he's decided not to limit himself to one denomination. In fact, what need has he for clergy at all? He's got a direct line to God now. But he wants to dip into other ecclesiastical waters because they interest him. The Margarets, baptised Catholics and in disguise as Protestants, are delighted to learn Pennsylvania welcomes Catholics. Well, welcomes might be too strong a word. Pennsylvania does not openly persecute Catholics, and theoretically the girls can come out of the closet here—but there is no Catholic church in the city. The nearest one is an adobe mission chapel in San Miguel (which will become Santa Fe) about 2000 miles west. Not only are Catholics generally despised (*Jacobites! Dirty Papists!*) in the colonies, aside from Pennsylvania they are not allowed to own land or vote. In fact, it's not a good idea to even speak Latin. Quakers and Jews are also given a hard time elsewhere in the colonies. Even the Church of England is considered insufficiently protestant, since the Reformation—it's accused of being too similar to Catholicism. In places like New York, the predominant Protestants have split themselves into dozens of quarrelling factions, with the biggest ones being the Baptists,

Puritans and Methodists. The Margarets are more cultural Catholics than spiritual ones, so their relief is not followed by an urge to buy rosary beads—more a sense of being welcome. They can be themselves. On hearing their Irish accents, one of the other Almshouse residents makes it known that rosary beads are available for sale, as well as vials of holy water and holy cards.

'How much?' asks my Margaret, her thieving fingers itching but told they *must not* pilfer here.

'What-a-ye got?' He's from Kerry. About her age, barefoot, face covered in chicken pox scars.

'Nothing.'

'A kiss then?'

'Sod off.'

'Alright,' he says with an agreeable smile and shuffles off good naturedly. Then over his shoulder says: 'Avoid the fish soup. It made me sick last week. And there's extra pillows in the box near the door.'

Margaret does not reply, but she begins to whistle softly—her way of thanking him, perhaps.

~~~

Things happen fast. By the end of the following day, which also happens to be his fifteenth birthday, David is working at the shipyards. His job is to do whatever he is told, primarily carry timber to and fro, and sometimes hold it firm while others fasten it to a framework. There's a lot of timber to move because each ship needs a minimum of 2000 trees—and it's heavy, for only oak is considered strong enough. Oak sap smells exactly like vinegar—not a pleasant smell initially, but David stops noticing because it's so pervasive. In fact, he grows to like it, in an unaware way. Vinegary air means everything is as it should be. The shipyard is a buzzy noisy place, but the heat and humidity sap his energy. He turns an alarming shade of purple, and his freckles multiply. Welsh skin has not evolved for sun, although evolution being what it is, some of his distant descendants will rarely burn. (Not me. I burn.) He has accommodation in the worker's hostel nearby. An airless space he shares with twenty-seven other men and boys, but from his top bunk he has a view of the port—and this suits him very well. He wakes early, finds the horizon of water, and commences his morning prayer. In silence, for he's learned the painful way that audible spirituality is not deemed manly here.

Meanwhile the girls are launderers in the Quaker workhouse, working twelve hours a day shoving dirty linen into huge boilers, standing on stools and stirring with long paddles. The heat is over 115 degrees some days, and their blouses are dripping by ten a.m. They're fed, housed, given the basics—clothes, shoes, some toiletries like hairbrushes and soap—but not paid. It's considered that their work is payment for their keep. The Quakers are kind enough, no one is beaten (in their presence), but the life is harsh. On top of this, there are the looks they get from men as they make their way back to the women's dorm after work and after eating. And sometimes hands grab at their skirt or blouse. Shrieking usually stops that malarky, but even so it's worrying. This is not how they'd envisioned their future in the exciting New World. They are drenched in disappointment, but in a tired stoical way. They do not weep or wail.

~~~

This status quo continues for three years. David with his developing muscles, darkening complexion and facial hair. The girls losing weight, getting their periods, and falling asleep the second their heads hit their chicken feather pillows. My Margaret accumulates a small hoard of pilfered items, none of them useful or valuable, and squirrels them away behind a loose panel. The seasons turn again and again and again. Back home, no one had told them the colonies had such extreme weather. Or they had, but they hadn't listened properly. Sun and snow can sound pretty in theory, even exotic when you're listening to constant rain on the roof. But that place—Ireland—seems so unreal now, like a dream. David visits them one day and is so shocked by their appearance—in particular Margaret Lyons, who looks two minutes from death—he proposes marriage. It's the only means of rescue he can come up with on the spur of the moment. Married workers at the shipyard are sometimes assigned minuscule rooms, and the wives become cleaners and launderers for the hostel. Only four such rooms exist in his hostel, but one of the husbands died a few days ago and his wife and children had to move out.

'Who are you proposing to?' ask the Margarets in unison.

'Either of you. Both?' he gallantly says, blushing. The truth is, it's only red-haired Margaret Lyons he has a soft spot for. Though currently her beauty is not apparent. She smells and looks quite unappealing. Even her hair has lost its vibrancy, there's an angry-looking sore on her lip, and she doesn't have the look of a person who loves to sing anymore. Her eyes are dead. Nevertheless, he clings to his original image of her feistiness and ploughs on.

During the week before the wedding, they bundle nightly—which is what couples intending to marry often do in the colonies—particularly in Pennsylvania—as part of courtship. A socially-sanctioned trial run at sexual attraction. With the Quaker Friend's approval—indeed, supervision—they both sleep in the tiny upstairs room in the Almshouse, the doll's house. The bed is slightly wider than a single bed with a small table next to it, on which sits a candle and a Bible. Keeping their clothes on, David and my Margaret—hardly able to look at each other—wriggle themselves into the bundling sack, a canvas bag sewn up the middle. And there they lie all night, but they hardly sleep for the proximity is arousing. Well, arousing for David anyway. Out of respect he tries not make his arousal obvious, or rock against her prone body—and the torture of this restraint is exquisite almost to the point of passing out. On the first night, Margaret is merely distracted. It's nice, for she's never been a one to enjoy sleeping alone—but it's too strange to have David right there, so close. Yes, she's had a little crush on him since she was seven, but it's been a chaste crush. The two whisper. Just inconsequential things. An annoying mosquito bite. The way the beef tasted a bit off earlier. They speculate about the Quaker woman who witnessed them getting into the bundling sack. Is this her house? Maybe she's the queen bee here, sleeping just below them. Or awake and listening out for impropriate noises. They become quiet, let their breathing slow and their limbs grow heavy. It's been a very long day and the moon has gone, leaving the room black. Then into this velvety darkness of non-consummation, comes David's whisper:

'Do you love me, Margaret? A little bit?'

His voice is so faint, she responds with a lips-closed question. *Mm?*

'Margaret, do you love me?' he repeats, slightly louder.

'Yes, I love you,' she answers. A lie! A lie of politeness, for who could say no to such a whisper from a betrothed? He has vulnerability in his voice. He is her familiar safe friend, not her lover—how can he be otherwise, given her age and their history? But a second later, as if her declaration has triggered some hormonal shift, has pulled love in—it's not a lie.

'I love you!' she whispers again, with feeling, and strains to be nearer him. Love, in her thirteen-year-old heart, is akin to a sudden rising wave of sadness, but one that is delicious and true-feeling.

After three nights of bundling, David is ready for the wedding night. Margaret less so, for she quite likes this stage. Of wanting and not having. The upside of sex is introducing itself, but only the faint beginnings. The dropping off of repulsion, mainly, and the surprising comfort of lying in the dark next to a person one likes. For different reasons, they're both exhausted by the end of bundling week.

~~~

On the following Tuesday David marries my Margaret in a Quaker ceremony. He is now nearly seventeen and she is nearly thirteen. Fine ages for marriage, mere months from being legal. They return to the shipyard hostel with Margaret Leary, who (they've decided) will for a short while be referred to as Margaret Lyons—sister of the married Margaret Williams and therefore eligible for residency in the barracks . (That they are sisters is the truth anyway, but no one knows that except you and me.) Hence, both girls are rescued from the workhouse and inappropriate attention from men.

~~~

It's 1772 and the idea of independence from Britain is well afoot here. Most of the colonialists are feeling confident. The ones who want rid of England call themselves Patriots and form into organised groups. War feels imminent and the colonies will win—ha ha! Silly old King George with all his silks and sulks will be crying into his porridge. David and the girls may not be politically inclined but feel a natural sympathy with the Patriots. And life is good. They are currently not starving or in danger. They've acquired the right clothes and shoes and habits to cope with the extreme weather. The newlyweds manage to lose their virginities a month after marriage, despite sharing their cubicle with the other Margaret, who sleeps on a mattress stored under the bed during the day. Now, when they have the time and inclination, they tie a ribbon to the outside door handle, and the unwed Margaret knows to take a walk for an hour. All is currently well in Williams-land. Yay!

Or not entirely yay. My Margaret is pregnant. Her body is too young perhaps, not developed enough, because she miscarries two months later. This is a relief, though also sad enough to warrant tears from both Margarets and prayers from David. This happens again three months later. No periods followed by a heavy painful period three months later. Many sighs, more prayers. Meanwhile the Margarets spend their days mopping, laundering and cooking with six other women, five of whom have children and infants stashed away in their rooms or working with them. Pregnant again just before her fourteenth birthday, Margaret gives birth to Elizabeth on May 30th, 1773. It's not an easy delivery, and at first Margaret turns away from her infant—not postnatal depression, more an adolescent pout. *How dare you hurt me so much! I'll never talk to you again!* The other Margaret tries to expose her friend's nipple to the crying infant.

'Stop it,' says the new mother, turned away, eyes closed. 'Go away.'

'Now Magpie, she's only hungry and wanting her mama. Turn around, darling.'

The argument goes on for five minutes, ten, fifteen. The crying becomes fainter. Finally my Margaret stops swatting her friend's hand away, sits up and allows access. Elizabeth objects for a moment, pulls away gagging, then plunges back and latches on vigorously. Suck, suck, suck. Cheeks pinkening as they pump the colostrum in.

'Huh,' says my Margaret, touched by how hard her baby is working. Then she says: 'Ouch.' But lets the child continue.

'Well done, Magpie.'

She has small breasts and hasn't given them much thought, but here they are, magically twice their normal size and fully functioning teats. It's like she's grown an udder.

'David, will you look at this?' she whispers. 'She knows what to do, like a wee kitten, so she is.'

David, who's been forgotten in the fuss of it all, looks at his young wife, his new daughter and his pseudo sister-in-law, all on the candlelit bed—a tableau of love.

'Thank you, God,' he fervently says, crosses himself and briefly looks upward. Part of him expects some sign again, but there is no discernible response from God. He turns his attention back to his family, goes to sit on the bed with them. It's almost summer and David's freckles stand out like spattered mud. There are black smudges below his eyes and a crease in the middle of his forehead—not a wrinkle yet, but it's always there. Despite his arm muscles, he still looks weak because he's short and built narrowly. His skeleton may keep growing for another year or so, but even if it does, he has the look of a permanent wimp. Not an obvious hero, and yet in another way don't you think his goodness, or his urge towards goodness, counteracts his physical disadvantages? It shines out of his eyes and in the consciously considerate way he moves. Here is humility embodied in a newly eighteen-year-old boy. And the Margarets—young enough to contain a wisp of freshness despite all the hardships. They're selfish compared to David, of course, and weak willed. Almost everyone around here is, and yet—aren't they beautiful too? And Elizabeth, who has turned this room into a nativity scene. She's so beautiful it's hard to look at her. Just a noisy sucking scrap made unwittingly by her teenage parents, with no expectations of the world beyond food and warmth. Then Elizabeth swallows air and starts crying again, and my Margaret cries too because the sucking has caused her uterus to further contract, and then the other Margaret starts crying because there are times when crying is contagious, but David can't help yawning despite his usual empathy because he's not slept in forty-seven hours. The moment of happiness passes with no one remarking on its rarity.

~~~

A year and a half later, just before Christmas, the same scene—only this time there is a toddler at the nativity. Sarah screams the house down from the second she's born.
'What's she so angry about?' my Margaret wonders.
'She's not angry, Magpie. Just hungry. Feed her, love.'
Margaret plugs her new baby on easily, having now become (at fifteen) an old hand at mothering.
Young Elizabeth sits on the foot of the bed playing with her ragdoll Sally, sneaking wary looks at her new sister every now and then. Sarah is not an efficient nurser and makes a lot of noise. Smack, suck, slurp, cough, grunt, wail. Again David is in the shadows, surveying the scene with something close to awe and thanking God out loud. Again there is no reply. No matter.

~~~

Three months later the war starts. Ta-da! David enlists in the Continental Army led by General George Washington. Almost all the men in the shipyard do the same. In fact men from all over the colonies are flocking to Philadelphia. Seamstresses organise themselves to quickly make uniforms—blue jackets, grey waistcoats, light canvas trousers—although mostly it's a rag-tailed army, with the men wearing their own clothes. They are recognisable as soldiers by tricorns and cocked hats, and by the long scabbards inside which swords nestle. Barracks are built, combat practice commences. Massive amounts of meat and beans and all sorts of edibles are purchased. General Washington is not crazy about the number of wives (or the number of women who insist they are wives) not to mention the number of children accumulating at the barracks. Dammit, what professional army has babies crawling around the place and women sashaying every which way? But it's all happening too quickly, and his main goal is to grow an army. Pretty soon he understands women can improve things. They cancel the need for hiring cooks and laundresses and cleaners. They keep the men relatively sober, well fed and clean.

The 372 wives and children are allocated a barracks of their own with two small rooms for conjugal visits.

The Redcoats make forays in their thousands. Win four major battles that summer and occupy New York, Washington and Boston. It's not looking good for the Patriots, who are far fewer in number and mostly untrained. But using guerrilla tactics, they gradually retake some occupied territories. Spain, France and the Netherlands start to send their armies, not because they love the colonies, but because they hate the British. This gathering of European powers behind the Patriots is great for morale and really pisses King George off. It's a very dramatic and dangerous period in history, a disaster for many—but it looks like the war is improving my ancestor's lives. The women's barracks means childcare is shared, something that did not happen in the shipyard hostel. And there's a real sense of comradery among the women—someone is always laughing or brushing someone else's hair. David's barracks is rough and there's no privacy, but when it's known he's a survivor of a shipwreck, and a married man with two children plus a resident sister-in-law, he's accorded proper respect for the first time in his life. Sometimes meek men draw the attention of bullies, but something in David's surety, his quiet dignity, is protecting him from that. Everyone likes it when he's around. All in all, war is a good time for David and the girls. They've never felt so safe, so full of purpose. So part of things.

The Margarets are washerwomen again, which is fine by them. Your mind's your own, when you're stirring boiling pots of dirty clothes all day, then wringing, drying and ironing them.

They're good at it. Quick and thorough, almost miraculous when it comes to stains, and people comment.

~~~

Let's draw back a minute. There they are, thirteen colonies clustered in the mid-northeast part of the North American continent. A bunch of Europeans choosing sides and killing each other, some of whom have already seen off the indigenous occupants with bullets and germs. Both sides have spokesmen who write eloquent articles justifying their stance, and their speeches are rousing and passionate—but all they are really saying is: Nah, nah, nah-nah nah! *This place belongs to me!* I'm reminded of little boys in sandpits, in an over-tired moment, waving spades and throwing sand in the eyes of other boys. *Go away!* It works, someone always leaves the sandpit in tears and goes running home, but the truth is that war, as a vehicle for change, is slow and messy. So why do we bother? Why does war, and sometimes genocide, seem an inevitable part of our destiny? Perhaps because Homo sapiens' predominant instinct is territorialism. Think about the Norse arriving in America in the tenth century—Leif Erikson leading the charge in the act of pillaging and raping up and down the east coast. And what about the original humans on the north American continent? The earliest evidence indicates people arriving from Siberia and Asia at least 13,000 years ago (possibly 20,000 years ago) over a land bridge now under the sea. These are the ancestors of the Cherokee, the Apache, the Chickasaw and almost 600 other native American tribes. The Siberian Asians may or may not have supplanted an existing society. Nobody knows yet, but if there were people in North America before them, it's likely they didn't fare well either. The newcomers wouldn't have taken the time to learn the language, sit down and discuss how to best share the resources. Humans are not good at sharing, but very skilled at justifying almost any act to achieve their aims. Think this doesn't apply to you and me because we hate war? What about that neighbourhood dispute about hedges and dog poop? And what about gun-toting land owners arguing over a fence-line judged to be one inch too far to the east? The real surprise is not that we wage war on the flimsiest of pretexts, but that there are not constant wars. The Finnish have a saying: Closeness without conflict is only possible in the cemetery. Some would say it's a miracle Homo sapiens are still around at all, given what a squabbling bunch we are.

But, to widen it out even further, it's worth noting that people are certainly not the only life form to wage war. Oh no, that's just gossip spread by self-loathing humans. Chimpanzees mark out territories and alpha male chimps take patrols along borders, killing and maiming, generally being terrorists. Other primates behave the same way, some even commit infanticide as an act of war. Termites go on suicide missions to protect their queen, and so do some worker ants. They blow themselves up, essentially. Cute little Meerkats have turf wars that make The Wire seem like a Disney series. Speaking of which, anyone who has seen Bambi knows that stags brutally fight each other for the right to the hind harem. Much seems to boil down to reproductive rights, as well as territory—perhaps they are the same thing. Fellow beings as territory one can own. Horses, bears, bats, giraffes, elephants and a million other species—they all commit atrocities against their own kind, as well as oust native species and destroy habitats in order to establish their boundaries. It ain't pretty, Life. I'm not excusing Hitler or Genghis Khan or the Vikings or indeed, the European settlers in America. Just saying the warring tendency is latent in all of us, and given the right circumstances, up it will rise. So curb your complacency.

~~~

Back to the revolution. It's January 1778 and Washington's army has moved eighteen miles north to Valley Forge to be safe from the British army, who're occupying Philadelphia. (Four British officers are now billeted in the upstairs bedroom in the Quaker Almshouse. One sleeps in the bed formerly used for bundling.) The snow drifts are twelve feet high in some places, and lie heavy on roofs of the two thousand rectangular log barracks built two months ago in a frenzy of construction. Winter had begun and most of the men didn't even have tents then. They dug holes in the snow to sleep in, pulling branches across the top of the holes. Generally, no one is whistling or laughing around here.

Washington is trying to convince his 12,305 men that he's not made a huge mistake, hunkering down here in the middle of winter. He needs to rally them, to re-train them, but he's having trouble keeping them alive. Food and medical supply chains keep breaking down, and there are other problems too. It's difficult to keep fires lit, and pneumonia, pleurisy and typhoid are killing men every day. Things are not looking good for David or the Margarets. David has become a scarecrow—albeit one that prays aloud until his fellow soldiers yell at him to shut the fuck up. (Yes, fuck is a swear now. It's been around a fuck of a long time, probably since the fifteenth century.) His wife Margaret has one advantage over her friend Margaret—she's tiny and doesn't require as much fuel to stay alive. Her third baby, my 4th great grandmother Nancy—arrived Christmas eve, the second baby to be born in the woman's barracks of Valley Forge.

Washington (who has no fantasies of becoming the first president of America, who nightly entertains the likelihood of defeat and imminent violent death—likely a public hanging at the hands of the British army) decides that in order to build the men's morale they need to take pride in their appearance. Surely it's harder to be disheartened when your uniform is clean and your belt is sparkling? The men are ordered to assemble in the central yard. Washington waits till they stop shuffling.

'Listen up! You will all take pride in your uniform from now on, and we will commence a daily morning parade where I'll personally inspect your uniforms for irregularities.'

Pause. The men look puzzled, two of them snort, and Washington edits his strategy.

'If you do not possess a uniform, that is acceptable. I still expect your appearance to be respectable. Men who fail to keep to military standards will lose a day's pay. Is that understood?'

'Yes, sir,' come the mumbles, despite the fact no one has received any pay. Anyway, there's no consistent currency in the colonies yet, so money itself is a nebulous entity. Things, big and small, are bought with units called pence, shillings and pounds—but these bear no physical resemblance to sterling. They're just pieces of paper and called Pennsylvania money, Massachusetts money, etc. They vary in value daily, according to the ever-changing reputations of colony-wealth. But this is not bothering the men today. They're too cold, hungry and scared.

'Stand to! I ask you again: Will you from today, keep yourselves respectable looking?' He summons the vestiges of his leadership tone for this command. It sounds false even to himself. But:

'Yes! Sir!' the men shout in rough unison, saluting and sending clouds of breath upwards.

~~~

My Margaret's Margaret, along with sixty-two other women, is assigned the infantry men's laundry. She collects the sodden stained clothes from the dirty clothes barrels, lugs them in a barrow to the wash house where a huge pot is on the boil like a witch's cauldron. She sprinkles them with de-lousing powder, scrubs them with lye soap, dries them on lines strung across the ceiling of the wash tent. It looks like a ghost army, if she happens to look up. All those empty dripping jackets and shirts and breeches swaying. The clothes never fully dry—it's too damp, even with the fires—and they are not ironed. When they've stopped dripping, she folds them and returns them in bundles to the men's quarters. It's a challenge remembering what clothes belong to which men, but she prides herself on getting it right most of the time. Their still-damp clothes smell of smoke and lye and of home. Washington was correct—clean clothes improve the infantry men's spirits and at least twenty-three fall in love with Margaret because of this returned sense of normality. Many more simply lust after her. One of them follows her back to the women's barracks one evening, pulls her into the woods and rapes her. She tries to scream, but her face is shoved into the snow and she struggles to breathe. Then, perhaps because she's been wondering when this would happen—for happen it almost certainly must, she's surrounded by more than 12,000 men without women—she submits and endures. After, he yanks her up, helps her sort her clothing out, then shuffles off into the dark. She waits a minute alone in the woods, noting the changed universe inside her. The painful rearrangement of things. Tries to recall the man's face, but that has already been wiped from her memory. It was a horrible act over which she had no control, but her subconscious (which is always looking for ways to aid survival) chooses to delete him. Then with the beginnings of a baby inside, she squares her shoulders and makes her way back. If she's lucky, there will still be some dinner left. Though on the other hand, what she really wants to do is go to bed and sleep. Maybe curl up with one of her friend's children for the comfort of another heart beating close to hers, and the warmth. The Valley Forge babies and children, all 692 of them, are passed round like parcels every evening. Are they mistreated, beaten, abused? Probably now and then, but no harm is recorded. At any rate, my Nancy will thrive well into her eighties.

~~~

Meanwhile, her mother, my Margaret (Mrs Williams) has been honoured with the job of laundering the officer's clothes—which include Washington's. Why her and only her? Maybe her thin face has acquired a chiselled beauty that flits into view often enough to warrant classier male notice. She can go from pretty to ugly by moving her head a millimetre, or from a cloud passing over the sun. Despite her recent pregnancy, her waist is back to seventeen inches. Her chin dimple is in its most attractive phase, as if it understands its role in keeping her alive is to make men want to kiss her. She's both fragile and buxom (breastfeeding again), a sexy look in 1778. David is proud of his wife. The officer's laundress! Other men might find an attractive successful wife threatening but not David. He's not typical, thank goodness for Margaret. This may or may not have a physical cause, like an underproduction of testosterone.

The other women tease her, saying she's all *la de da* now with the officers, a snob, but she doesn't care. She teases them back, saying she pities them in their lower-class pit of filth, but the truth is—dirty clothes are dirty clothes, and she only finds the usual range of nonsense in their pockets, hardly worth stealing. Though one day she comes upon a gold sovereign that has slipped into the lining of an officer's jacket pocket. At last! A prize worth keeping! The profile of the recently dead King George II winks at her encouragingly, and into her bosom the coin goes where it nestles until it joins her other loot that evening. A week later, she sews the coin into the lower band of her corset, where it remains until discovered by a woman called Bethany in 1835. Hidden away, the never-spent coin is essentially pointless—unless it provides Margaret with a vital spark of self-worth. Anytime she begins to feel small, maybe she reminds herself she's a woman with gold on her person. Meanwhile her hands are red raw from soap and boiling water, and her back aches from carrying heavy baskets of laundry to and fro.

~~~

Factual diversion: Margaret's great grandson Sam will marry Cecelia Mahnkey, a laundress of great skill in Redding, California. And I realise it's probably just coincidence that I felt at home when I became a laundress, aged twenty-two, at the gas workers' dormitories in St Fergus, Scotland. But seriously—there might be a conflation of laundering genes in my DNA. I've never found washing clothes to be a chore. Cleaning bathrooms and mopping floors, on the other hand…not so fun. Biological imperative? Perhaps a lower chance of suffering from conditions contracted through filthy clothes. Can you get sepsis from over-scratched flea bites? Or maybe Washington was right, and clean clothes simply produce a sense of well-being. Illusory or not, contentment can support your immune system. Think of getting into bed when the sheets have been changed. Nice right?

~~~

George Washington's shirts are long white linen garments, always soiled. Underwear has not arrived yet, and every morning he simply tucks the long shirt tails between his legs like a diaper before putting on his stockings and breeches. Sometimes Margaret has to scrub off a brown smear on his shirt tails. This can be nothing other than the excrement of General George Washington. It tickles her. Their leader shits! Not only that, he sometimes fails to clean himself properly. One day when she's returning ironed clothes to the officer's quarters, she runs into Washington in the yard. They're both stepping onto precarious planks through the slush.

'Good afternoon, Miss,' he says.

There's no American accent yet, not in the sense there is now. He's a third generation American with English heritage—these things matter, in a class-sense, and he feels superior to new immigrants just like everyone else here. English accents have already begun to change with exposure to so many other accents. It's not just interbreeding that results in a new ethnicity (or sometimes a new sub species, like mules being the product of a donkey and horse. And don't forget ligers, the products of lions and tigers!), it's unconsciously mimicking other voices. But still, men from families like Washington's tend to clip their t's and avoid pronouncing r's in the upper class British way. He bows slightly now to my Margaret. I think it's a mock subservient bow—or maybe not, because his cheeks, which are always pink, pinken further. Does he think she's pretty? (Imagine if *he'd* become my 5th great grandfather, not David. No, that never would've happened. Attraction to powerful charismatic men is not my Margaret way.)

'Is it a good afternoon?' challenges Margaret cheekily, her accent straight from rural Ireland, her basket tilted on one hip. It's a fallacy that respectable wives with three children aren't flirty. Heavens! Look at her quivering cleavage, exposed even in this weather.

He stops. Looks up at the leaden sky and, shrugging his shoulders, says crisply though perhaps also a mite shyly: 'No. Indeed it is not. I lied.'

'So you did. It's a shite day, sir. Another one.'

It begins to rain, as if on cue. Then it's sleeting. Some of the sleet is snow. Margaret smiles broadly now, lets her face be sleeted on without so much as a flinch. Her smile could be described as brazen. He wipes the wet off his face and smiles back despite himself because he's suddenly seeing this scene as if he's not in it. How it would make him laugh, if he was over there watching. He thinks that if anyone *is* watching, he'll need to reprimand this insolent washerwoman. He glances around quickly—no one there, so he laughs instead. At first it's a light chuckle, then it grows into a deep belly laugh. By God, he can't remember the last time he laughed like this. He almost falls off the plank. Nothing remotely this funny has occurred. It's giddiness born of exhaustion and strain and the presence of a cheeky Irish girl. Entirely inappropriate in a General responsible for the lives of thousands and the establishment of a fledgling country.

Margaret freezes, then laughs too. For a full second, maybe two, something warm passes between them like an airborne elixir. George is forty-five and long married to Martha, though there's speculation it's not a real marriage, that he married his wife for her 18,000 acres and 300 slaves. Yes, of course he has slaves. Almost everyone here with a bit of money has slaves. In fact, there are slaves right here in Valley Forge—some owned and without choice, others freed.

Washington has just been flirted with by a washerwoman and this makes his heart pump faster, and some swelling of the loins occurs—which surprises him. That's not happened in a while, arousal being a calorie-burning event not affordable in times of imminent oblivion. He and his wife have no children, aside from the two she has from her previous very wealthy husband. (Clearly Martha doesn't go for under-achievers.) As for my Margaret, the idea of a relationship or even a flirtation between herself and General Washington is absurd, but still the moment, this one we've fallen into (an elongated time frame in which the normal rules don't apply) can't be denied. They pass each other on their separate ways, each with a smile tucked into their cheeks. For him, it's an imagined kiss, and more. Removing undergarments, pulling down bodices, firm buttocks. For her, an imagined introduction into higher society, her spouse and children having momentarily evaporated. Hey, everyone dreams now and then. Fantasy has a very important biological function. At the least, imagining alternate realities temporarily alleviates stress, hence fortifying the immune system and lowering blood pressure. At the most, fantasizing in a negative way (imagining dire outcomes) can increase chances of surviving risky situations. Don't walk alone down that scary street in the dark, desperately bad people with knives are hiding in the shadows!

~~~

Life trudges on. Dramatic military scenarios are being played out all over the colonies, but here in the fort it's a matter of eking out food and not freezing to death. In early May, after the thaw, my Margaret's Margaret marries John Beaton, one of the soldiers who's been in love with her since they arrived in Valley Forge. He's agreed to claim the child (when it comes) as his. She's so grateful she almost loves him back, although not enough to fancy him. My Margaret, her David and their three children continue not dying, which is lucky. By June, when Washington and his troops (and their wives and children) depart Valley Forge to set up in re-taken Philadelphia, over 2000 soldiers have died of disease, malnutrition, and the cold. It may have been 3000, had their uniforms not been so clean, or it may have been exactly the same. The dead—women and children included—are buried in unmarked graves on the outskirts of the fort. Being winter and the ground frozen, the graves are shallow and square, bodies interned in curled up positions. As the troops march past the graves on their way south, some salute the ones they are leaving behind.

~~~

David is a good soldier, considering he puts spiders out of the barracks instead of killing them. He fights in five battles over the next four years. Each time, in the moments before combat—that vast frozen period when the whole ragtag army seems to hold its breath—David is afraid. His bowels loosen, he hyperventilates. He feels dizzy but also experiences an adrenalin-charged clarity. There they are, right in front of him! Not just the enemy in terrifying numbers, but their presence reduced to the things he's terrified of—of killing, of pain, of dying. It seems impossible that he'll submit to one or all of these fates in a minute. That his muscles and mind will cooperate enough to move from this frozen position. And then come the moments after the call to action. Of doing the impossible things. An immense sense of unreality descends as if he's watching himself from afar. And weirdly, it becomes almost easy to be part of the battle—for, second by second, it is never as frightening as anticipated. It is just another stretch of lived time in which, yes, terrible things are happening—but also all the normal things which must occur no matter what. It's all mixed up, the ordinary and the extraordinary. Breathing, blinking, moving one's feet, moving one's whole body, feeling annoyance when falling, noticing a need to defecate, hoping one does not, noticing a friend is now lacking a leg, hearing a British soldier and noting a familiar accent—is that man Welsh? Recalling how good the beef tasted last night, just the right side of rare. Irritation that his ears suddenly aren't working, that sound has become muted below a loud rushing wind seemingly generated inside himself. A sense of deja vu, with this deafness, and then remembering previous battles and thinking it odd that he keeps forgetting this about them: That he continues to be himself, even while doing the unthinkable. That wondering about an accent can co-exist with the effort required to shove his bayonet into another man's chest. Is killing a sin when a man's actions comprise automatic and instinctive reflexes? To stay alive for his wife and children is crucial, and when he kills a man, he also prays for him—sometimes a second later, but mostly when he can process the day's violence calmly. *Dear Lord, please welcome another soldier. Black beard, rather fat. Pox scars, broken nose. He was like me and only wanted to stay alive, and now he's not. Please forgive me for taking his life. And thank you for allowing me to continue not dying. Comfort the dead soldier's family.* On and on.

~~~

Meanwhile the babies keep coming. In the four years until the end of the war, David and Margaret have three more children. A girl, Ellen, followed by two boys—sons at last! The first is called William, of course. The second is Daniel Nathan, a good solid name for a first generation American. Six children by the time David is twenty-six and Margaret is twenty-two.

~~~

It's 3 September, 1783 and the war is over. Yesterday they lived in the United Colonies and today they're in the United States. The bad news: the end of the war effectively means David is out of work and a place to live. The good news: he's given an Ohio land grant in recognition of his part in the revolution. This echoes the original distribution of land in the United Kingdom. Tracts of valuable land as reward for loyal service in battle because territory (of course) wrested from opponents immediately requires ownership by trusted compatriots. My Margaret is delighted, for the land grant makes David a man of property and that is aphrodisiac enough for any girl. She's been having marital sex for almost ten years now, but rarely with enjoyment. Too young when wed, then too tired from motherhood—and anyway, who expects a woman to like sex? She's certainly never witnessed a farm animal enjoy being mounted. Even the hens scurry from the rampaging cockerel. Now, for the first time, she fancies her husband. When he kisses her, her skeleton melts. Lust feels like a delicious malady. She confides in her friend, but it turns out Margaret Beaton is only having dutiful sex with her own husband, so my Margaret tactfully ceases the confidences. Instead she says:

'Come with us to Ohio, Margaret. It'll be great.'

'Ohio! Huh!' She pronounces it *Oh*-hee-oh. 'Head west in some dirty old wagon, hope not to get scalped by the Iroquois or lost in the dessert or eaten by wolves? And then what, Magpie?'

'Then we build houses, plant crops and live happily ever after, silly billy!'

'Also, not enticing.'

'And you'd keep me company and generally save my life.'

'Stop exaggerating.'

'If you don't come, I'll die.'

Pause full of Margaret Beaton's sighs.

'Hen's teeth, when you put it that way…'

~~~

The prospect of living hundreds of miles from even the slightest vestige of civilisation—these things would worry the Margarets, if they were not going together. The plan is for their husbands to work land adjacent to each other. (John Beaton had been given a Missouri land grant, but managed to switch it for an Ohio land grant.) To share farming equipment and breeding stock and crop seed, for their children to grow up together. Everyone has a fantasy of the future, and off they go one early spring day, pots and tools swinging from their covered wagons, babies and children hanging out the back. It's a three-month trip west, beset (as predicted) by scary Iroquoians and Crows with their hunting bows and knives, by wind-whipped sand, by lack of clean water and fresh food. The men drive their own wagons, and after a while the women take turns riding with each other and the children—six Williams's and three Beatons. It's just too lonely otherwise, for the men are not much for talking. John's conversation is limited to: *Them beans ready? Where's ma boots? I left 'em right here.* David prefers silence so his ongoing conversation with God can continue unabated. More and more he has a permanent look of vagueness, as if he's elsewhere. The women are close, but the men too dissimilar—and in any case, neither see the need for closeness. Friendship feels a feminine thing. One night John drinks himself almost to oblivion, falls onto the marital bed (the thin horsehair mattress that gets unrolled every night), fumbles with various of his wife's body parts while babies and children snore nearby, and the deed is accomplished. *Go sperm! Not that way, this way!* Not a particularly romantic or passionate moment, but thank God! My fourth grandmother's saviour has come into being. Whew, again, and hooray for tiny Seamus! For Nancy's life will be saved by Seamus no less than three times, allowing her to reach reproductive age in fine fettle. So in theory I owe my existence to him too, proof that having children is not the only way to preserve your species. Meanwhile my Margaret is pregnant again too, about two months and taking it in her stride. Like a brood mare in her prime.

~~~

The two families settle in the territory of south-east Ohio in what will become Wayne County. They begin the rest of their lives. David and Margaret are in the place they will end up dying in fifty-one and forty-seven years respectively. Margaret Beaton and her husband will likewise breathe their last here, though the offspring of both families—sixteen children, eight of whom sport chin dimples—will end up all over the state, only four remaining in Wayne County.

The adventures aren't over, but the way they occur in time will be different. When David and Margaret look back, the years prior to Ohio will seem to take up almost all their lives, crammed with memories. Moving to Ohio will forever feel like a recent event. A compressed period in which much happens but little is remembered. David and the Margarets will never become mid-westerners. They are Irish and Welsh to the bone, and just happen to be in Ohio. Far flung flukes.

~~~

Laws are made and documents signed in a smoky Philadelphia office by fourteen bickering men who call themselves the Continental Congress. This title works alliteratively, but half the men think it's not sufficiently weighty, so after a while they vote to call themselves The Congress of the Confederation. Regardless of those earnest men or their carefully worded documents, laws are rarely enforced out west where the Margarets and their husbands are. There's a sense of lawlessness here, a lack of safety, of being held to account, which is marginally mitigated by the sheer distance between houses. If you can see a stranger coming a mile away—why, you've got plenty time to grab your rifle. There is no nearby town as such, no post office or school or sheriff's office or general store, but there's a cluster of houses about eight miles east. In a few years this will expand into Waynesville—named after General 'Mad' Anthony Wayne, a Quaker with charisma, and for whom David will have complicated feelings due to Wayne's cruelty to his many slaves plus his numerous marital infidelities. A Quaker with this lifestyle is deeply disturbing. They will never meet, David and Anthony Wayne, which is probably a good thing.

~~~

No Williams babies are born in Ohio, unless you count the one who nearly kills his mother that first autumn. The house, not even half built. What makes a couple of farm kids turned soldier and laundress think building a house is something they can do? The cold winds have begun, and Margaret's pains come fast but lead to nothing. The six children keep quiet at a distance in the shadows. David paces and prays. Hours and hours of contractions, and the spells in between growing shorter. Margaret's face is grey, the pallor of a dying woman. For a good three hours it seems God is going to desert David after all. To take away his wife, the mother of his children. Where's the logic in that, the fairness? To save her from shipwreck, just to let her die in this agony? But he keeps praying, keeps up the running conversation, and then she appears at the door. (This counts as the third miracle.) A woman, maybe Chippewa by her clothes and facial features, not young and not old, dressed in rich blue and gold and wearing exquisite moccasins. Hair and eyes, black as coal. She doesn't smile, doesn't speak. Just lets herself in the room and moves to Margaret, prone on the bed. She nods once to the children, in a friendly way, then waves them away. She puts her hands on Margaret's belly, just lays them there for a few minutes, then rolls Margaret to her side and feels her belly again. Begins a low singsong humming while rocking on her heels. With sign language, indicates to David to boil water. To build up the fire.

'Is she going to die?' he asks in sign language, still thinking *Who on earth are you?* She looks vaguely familiar—maybe she sold them some blankets? He met some Chippewa men when he purchased livestock. They'd seemed alright. Not scary, like the Iroquois. Not bitter, considering they'd lost their land to the likes of folk like himself. But how did she know to visit this half-built house tonight?

The Chippewa woman speaks some words he doesn't understand, then shakes her head and shrugs at the same time. Not reassuring. She takes some dried plants out of her pouch, lights them on the fire and wafts the smoke towards Margaret. Sage and other sharper smells David does not recognise. The wind picks up and whistles, finding all the places where wind can sneak inside. All the while, the Chippewa woman keeps humming and sometimes crooning. Margaret closes her eyes, goes limp, and then—eyes still closed—gives a guttural feral noise, the like of which David has never heard during her other labours. She's on her hands and knees now, her face scrunched up. She repeats the noise at intervals, and finally the baby's head crowns. Another throaty grunt, and the body slithers out into the Chippewa's hands. The infant does not cry because he has no breath in his body. Not a hope of breath, he's been dead for some time, and the Chippewa woman wraps him in a cloth and hands him to David. But there is no time to mourn, for Margaret—still pale, limp again, on her back—has begun to haemorrhage. The mattress becomes soaked and the room fills with a rusty tang. The children begin to whimper and the younger ones to cry. The Chippewa woman doesn't frown or alter her expression in any way—but she moves quickly. Her hand disappears into Margaret and tugs away the placenta, slings it to the floor—all the while, her other hand is massaging Margaret's stomach.

'Sugar! Water!' she barks, her first English words, and David jumps.

Then he complies and she indicates he should rouse his wife. Try to rehydrate her. The stomach massaging continues, more blood pours—but less so. Within a half hour, Margaret opens her eyes. The children rush to her and the Chippewa woman allows it, before departing the house as quietly as she entered it. Margaret's recovery takes many months, and David sleeps on a mat by the stove. They have different feelings about this arrangement. Three years pass, and she still feels rejected and he still feels noble. Now and then, separately, they each remember the eroticism of bundling—sleeping with a barrier to prevent intercourse. That pre-martial week in the Quaker Almshouse. Sadly, it doesn't occur to either of them to reinstate it.

~~~

Thank God for God who, now more than ever, is the third person in their marriage. It's been a long time since David felt the need of a church or minister, which is lucky, as there are none here. The solace that comes from his continuing conversation with him (Him) saves him over and over. God has not spoken out loud since the ship capsizing, but that once seems to be sufficient. He recalls the voice precisely—the timbre, the volume, the urgency. The sense of familiarity, of intimacy. The voice of a speaker who knew him inside and out, and loved him. So, what are these one-sided conversations about? The weather. Whether or not it's alright to enjoy listening to his wife's gossip. The health of his family. Should he plant the seeds now? Harvest the wheat now? When to kill the hogs or let the bull into the cow field. Is his wife alright, and if not, is it his fault? Are his older children working too hard? Hard enough? Should the girls be taught to read? Should he loan John Beaton the plough even though the last time he borrowed it, he returned it with a broken blade? *Please let Nancy's pet lamb not die. Please let William find the pocket knife he lost while hoeing the bottom acre. Please let the sun shine. Please let it rain. Please let the snow stop. Bless me, bless this house and everyone in it, bless the roof and keep it from leaking. Bless this farm and all the animals. Bless the corn seeds in the ground and the peach trees and the berries. Bless everyone who has died, especially Mum and Dad and my brothers and sisters. Bless my grandparents and all my ancestors. Bless everyone back in Wales, and everyone back in Ireland. Bless the turkeys and buffalos and deer and elk and raccoons I've killed and eaten. Bless the men I killed during the war, may their souls rest in eternal light. Bless their families, their children and their children's children.* A long litany of blessings. He often rocks gently during this phase of the conversation, for the rhythm of it. He's not aware of doing so.

~~~

Meanwhile the Margarets, who are only mere acquaintances of God, see each other as often as they can. It takes almost an hour to walk to each other's houses. Unless there's a blizzard, they manage it. They bring the younger children of course, but not the men, who in any case are always out in the fields working. No matter what calamity has happened during the week, the women find time to sit in the kitchen drinking coffee (much more patriotic than tea) and unpack it. Turn it over and over until it's something manageable. An anecdote. A misunderstanding. A loss from which some small thing can be salvaged. Then, before dusk, the visiting woman gathers her children and sets off for home. It's no exaggeration to say these visits are vital to survival. On some individuals, isolation can inflict as much damage as starvation.

~~~

David and Margaret are too tired every day to notice how the horizon is affecting them deep down. The dome-ness of the plains sky, the flatness of the land. They grew up by lakes and the sea, with wooded hills all around, as did their ancestors back countless generations. The sight and smell and sound (yes sound is quite different depending on where it occurs) of those places went into the making of them, and Ohio can never be their natural habitat. Without realising they're doing it, their breaths become slightly more rapid and shallow, and their sleep less restful. They'll never know we're all genetically predisposed to prefer certain landscapes. Some people are natural hill dwellers and require views to feel right, some are intrinsically landbound and feel ill at ease by the coast, and some are valley folk who cannot abide open spaces or views from heights. They need to nestle in, hide away, and don't feel fine until they do. People who are not in their right place tend to feel uneasy. A chronic low grade edgy feeling, as if they're waiting. But David and Margaret are adaptable—it might be their key strength—and eventually they accommodate this particular (there are many) sense of wrongness. It doesn't go away, but it's managed.

~~~

On her thirtieth birthday, my Margaret learns some news that is weeks old—George Washington has become the first president of the United States! He's now a superstar. She immediately reminds her family that she used to wash his clothes. It's the only claim to fame she has and she milks it shamelessly in kitchens, in the new general store, on sidewalks to strangers passing by during her quarterly trip to the fledgling town of Waynesville. But she never talks about Washington's dirty shirt tails, those brown smears. About her image of the great man squatting in the bushes, then incompetently wiping his bottom with some leaves. That, and the bleak Valley Forge afternoon she made him laugh by contradicting him. *The afternoon is shite, sir*. These remain private, not to be aired in public. *My George*, she thinks sometimes if she's feeling sorry for herself. He'd never drag her off to a place like this. No, no, no! Her George would hire servants to do her bidding. Escort her to high society balls. But that soon fades. He was an old man! Marriage would've meant catering to an old man's fussy ways.

David has an entirely different view of Washington, having been on five battlefields with him. He remembers being surprised how ordinary he seemed. A leader who sometimes picked his nose while waiting for his men to fall into line. Who seemed to lack the confidence or the ire to punish insubordination consistently. Who, now and then, was indecisive when decisiveness was most needed. Once he saw him leap in the air and scream like a girl when a cannon ball hit a nearby boulder, making sparks. Even so, with all these proofs of his common humanity, Washington was impressive. David remembers being wounded at Monmouth, shot in the shoulder by a British musket and flung to the ground. When he came to, having been left for dead under bodies of soldiers, he found himself scooped up in Washington's arms. Washington was a sturdy 6'2" to David's frail 5'5", so it was like being a child in a father's arms. It didn't matter that Washington wasn't his father. He'd been a father when a father was needed—and maybe that's what David had always wanted and still does. To feel safe, protected, not alone. His own father, William Williams, had been a good man, but there'd been too many children for him to notice one of his sons yearned for proper attention. For someone, anyone, to fill the father role. Can you have too many fathers? Apparently not. Later, bandaged up, he thanked Washington, who had no recollection but who was gracious enough to smile. Give a brief salute and say *You're welcome, son.*

'Good news, about Washington. Good man,' David says that night on his way to his boxroom bedroom.

'Darn tootin,' calls Margaret from the marital bed, where she's snuggled under five blankets with their two youngest. Nature hates a vacuum, and since David's nocturnal departure she's only slept alone five times.

~~~

How affected is David by these years of no sex? He's still young, in his thirties. He doesn't become mean to his wife or begin affairs or seek out the one prostitute in Waynesville. Nor does he commit himself to thrice weekly attendances at the new Quaker meeting house. No, no, not my good and odd David Williams! What does happen, and which may or may not be connected to sexual frustration, is this: One day, a market day after he's bought and sold various cattle and hogs, John Beaton suggests they visit the saloon. They're invited to join a game of poker, so they do because it seems the polite response. David wins the pot, which is $39—a fortune. He's never played cards, doesn't even drink—is drinking lemonade right now—but he falls rapidly into the vice of gambling anyway. He returns home late that night with empty pockets. It's an old story. One he confides to God daily, sometimes several times daily. How did this happen? He does nothing but try to be good from sun up to sundown, and yet here it is—sin has wriggled its way into his being. Not only is it wicked and impoverishing, it's also a distraction from his God-conversation. Nevertheless, it's committed over and over for nine years till the farm is in hock and his wife hardly speaking to him. But then, isn't it her fault for not being willing to risk pregnancy again? Or is it the doomed seventh baby's fault? Or David's fault for impregnating her? Or Captain Mack's fault for getting so drunk he failed to take the hurricane seriously early enough? Or the sore tooth in Captain's Mack's mouth's fault? Or the sugar cane he loved to suck when young, which rotted the tooth? Nothing good comes from appropriating blame. Back and back it goes until you could end up blaming, well, if you are like David on a very low day during this phase—blaming God for making the world. Such a rotten to the core place!

Oh, David. Oh, Margaret. And things were going so well!

~~~

Meanwhile Margaret's friend Margaret Beaton endures her endless round of farm and domestic chores, as well as offhand cruelty from her husband. Over cups of coffee, she tells my Margaret the truth about her marriage. She now clearly understands the deceitfulness of love, and why did any woman desire marriage? It's obviously a trap, a trick. The social stigma of unmarried motherhood had driven her into John Beaton's arms—but he'd not seemed a bad man at first. He'd seemed grateful, in fact, and his kisses had been so sweet. Her rescuer! Those kisses promising respectability, plus her pregnancy hormones, had heightened her senses but sedated her brain. Romance was a trick to make sure women married and had babies, of which she's had seven now and only two in the ground (well done!).

'It's all a huge humbug.'

'But real love also exists between men and women,' insists my Margaret. Currently she has no evidence for this view, aside from the longevity of her marriage—and that's no evidence at all, for what choice do either of them have out here on the plains with hardly a penny to their name because her husband is a gambler, and six kids to keep alive? Maybe she's just in a contradictory mood. Or maybe she's her mother's daughter after all, insisting on an optimistic view of the world. She's stirring a pot of berries into jam, and her long curly red hair is pulled back into a bun the size of a baby pumpkin. Her arms are still like sticks, and her work-worn hands like claws.

'Does it, Magpie? You sure about that?' Margaret Beaton replies, her eyes narrowing. She's not in the mood for being cheered up. Or for being told what she's missing out on. What she requires today is affirmation. For someone to say: *Yes! Romantic love is just lust, and lust is just nature's way of tricking a girl into having babies!*

Well, I'm asking you—is it? Maybe it is. But what if the biological imperative of romantic love is not just about babies, but keeping a beloved alive. Keeping oneself alive too, for the act of loving can engender love—and it may be the case that beloveds live longer. Lust can, at least temporarily, lead to a mutual protection scheme whereby the lovers instinctively—without thinking for a millisecond—throw themselves in harm's way to save a beloved. A lucky few find romantic love and the longer kind of love in the same package with marriage, but probably not many. Not Margaret Beaton. Not my Margaret, now that her husband doesn't touch her and has become a diehard gambler. But there are alternatives to marital love. Friendship, for one. It's always tickled them, that they're both Margaret. *The marvellous Margarets.* They dress for each other's pleasure when they meet, although not consciously. My Margaret often wears her low-necked navy linen robe over a cream petticoat, and Margaret Beaton (quite stout now) favours a yellow cotton robe, roller-printed in a yellow rose pattern with a bodice that allows her bosom to breathe. Dresses are not here yet—robes, or gowns, are worn over petticoats, under which lie their shifts and corsets. Between them they possess seven outfits, and consider that quite sufficient. Look at them, sitting at that kitchen table—aren't they pretty? If they'd been in another generation and place, perhaps something sensual might've grown between them—or maybe not. All that matters is that their love is nourishing, almost medicinal. The Margarets don't wonder much about their attachment, whether it's normal or not. In any case, it's too omnipresent to consider objectively. This conversational attack and defence of love is exceptional. Mostly they are not reflective women, too distracted by what every day throws up to them.

~~~

David, on the other hand, lives much inside his head and has considered love from many angles. He's come to various conclusions. Some days he thinks love is internally generated and specific to each individual. Custom-made by effort to suit the chosen beloved. Other days he's convinced love is something existing outside himself, like a sparkling stream he cannot see, or a sudden patch of warm air on a chilly day. He's tapped into it now and then—but accidentally of course, for no one wakes up one morning deciding this is the day they'll lasso love. Some days he wonders if it's simpler than both these ideas. Maybe love is God, and that's all there is to it. As John says in the Bible: *God is love, and all who live in love, live in God, and God lives in them.* David can still summon the memory of God's voice, much as a person might pull out an old photo, creased and worn, and try to relive a first kiss. He can still replay God's voice precisely, and at these times David briefly feels the membrane between himself and God thin out. The distinction between them is not a sharp outline. It's not that David thinks he's God—it's that, like some Quakers, he thinks God might be in everyone. That he (He) is right inside a person from birth, observing without directing. Very rarely issuing advice. A benevolent hitchhiker.

~~~

Ohio becomes a state on March 1st, 1803, the same day David (forty-eight now) has a short-lived epiphany while milking by lantern light at dawn. It may be spring, but they had five inches of snow yesterday. On the prairie there's only a few seconds between pitch black and full sunlight, so by the time his epiphany is fully born, the lamp is redundant and the day has begun. He's been so angry at himself lately, for promising God not to gamble and then gambling again. He's become as selfish and childish as a…as a selfish child. The self-loathing goes around and around. His head is pressed against the beast's heaving side, his fingers squeezing the hot milk down through the teats and into the metal bucket. Squirt! Squirt! Both his and the cow's exhalations are visible. David likes milking, always has, and he likes this particular cow—Rosie. He feels her relief as her udder softens and empties. He easily imagines being her. And then he thinks: But of course it's easy to slip into a cow's consciousness, for aren't we animals too? Not only that, we're also very young animals. Playing dress-up, aping maturity, and this suddenly explains everything. War, lust, weaknesses and foibles of all kinds. *Gambling.* This is a crystal-clear realisation, a vision. His milking done, he crosses himself and credits God. *Why, that's what the problem is! We're all about four years old. Having tantrums, making wishes, worrying someone will know we still wet the bed.* He thinks of the different ways each of his children has, on occasion, disappointed him—a peculiar complex pain he'd not anticipated with parenthood. What is wrong with these kids? So lucky, the first-generation Americans, compared to his and Margaret's experiences. Daniel and William are still not married; do they think their mother will wash their clothes forever? And the daughters, all married, take turns causing anxiety—one sporting black eyes every month or so, one who cannot keep her children's faces clean and their clothes mended, one who never visits and hardly offers a cup of coffee when visited, a teenage grandson with the intelligence of a baby, another grandchild who cries too much. It's suddenly easy to understand a creator loving each of his creations no matter how awful they are. To sometimes cringe when they misbehave, but then tsk and shrug and look away. *After all, they have free will and they're just kids! Of course they're going to wreck the place.* Then he's kneeling on the ground, in the straw and muck, head bowed and rocking slightly. Forgive me. Forgive us all in all the places we've sinned—the saloons, bedrooms, battlegrounds, marketplaces, sailing ships, cobblestoned lanes on dark nights, wide open prairie spaces. Forgive us when we sin because we think no one is watching. Forgive us when we think we are sinless and not needing forgiveness. Forgive us the sins we have committed in our hearts, thinking they do not count. Original sin is deep within us all, God, and even the saints among us are tainted. We cannot help but sin.

I'm sorry, I'm sorry, I'm sorry. Mea culpa.

I'm sorry and I'm sorry, and I'm also sorry.

He swells with tranquil humility. His eyes calm and his

gait easy, he brings the milk into the house. Spills some on the floor and Margaret shouts that she's only just finished mopping, and why does he have to be so damn clumsy? His epiphany starts to slither towards the ether—what's her problem? It's not like she's flawless, with her petty thievery filling the house with dumb stuff. But then he presses his lips firmly and commands the epiphany back. *Forgive her. She's a child, I'm a child. We're children raising children in a universe run by children. Amateurs, all.* It works until he catches her scowl at his inept attempt to mop the milk. Perhaps his gambling has depleted not only their bank account, but her respect. Made her mean and ugly around him. Not good for her, not good for him. Both slump inside and feel heavy, trapped. Turn away from each other and move on with their day. Even the epiphany cannot help David through times like this. The unhappiness is simply too big.

~~~

I so want to believe that as a species, we're always improving. That evolution is a refining process which happens on a genetic as well as social level—but that could be illusion. Like the way hoover manufacturers come up with a new improved product every single year. *Surveys prove 99% consumers agree this is the best hoover ever!* Maybe they just think we won't buy it if it doesn't claim to be better than the last one. But it usually sucks up roughly the same amount of dirt as before. At least in my house.

~~~

David and Margaret enter old age. Sometimes close together, mostly parallel to each other, each with their own concerns. The gambling bug burns itself out one summer's day as mysteriously as it arose. Is gambling a virus like measles? Can it be caught, and if not fatal, shrugged off later? Are there antibodies that slowly defeat addictive disorders? David wakes from a nightmare. How could he have been so stupid and weak? What on earth had been the appeal of playing cards? He greets his old sensible self like a dear friend he's not seen in decades, had never expected to see again. Someone he'd assumed was dead. *So wonderful to see you again! I hope you can stay for a while this time.*

One morning not much later, over breakfast, Margaret notices David no longer has freckles on his face. Where have they gone, those characteristic dark specks cluttering his features, and how has she not noticed their absence until now? Other things have faded too, and not been noticed until absent. Margaret's tendency to sing while working, and her feistiness. Her light-fingeredness. David's sunniness and carefree way of swinging his arms as he walks. Gone, gone, gone in daily increments too tiny to notice, ground down by the minutia of life's problems. But some things are better away, and the absence of gambling is something to celebrate. Margaret takes credit, for hasn't she been sulking for thirty-three years to achieve this end? David, of course, credits God. It's a miracle. (The fourth.)

~~~

And yet. And yet! Something is happening with God. It must be, for around the time of his seventieth birthday David feels an absence, a chilly void where once there was constant company. Like a bat noticing free air space by the changing quality of an echo, his mind's eye reaches for God and draws a blank. *Emergency, emergency!* he wants to run around shouting. *God is missing!* Of course, it's possible that God has been an imaginary friend. That David was a lonely child, and his need pulled Him into existence. It is a scientifically accepted fact that even adults sometimes hallucinate the things they need. For instance, Charles Bonnet hallucinations. Partially-sighted isolated people vividly seeing things that don't exist, in order to stay mentally healthy. Just ordinary things, usually. A friendly mutt to keep the black dog at bay, or a familiar friend or long dead spouse or parent. Sometimes a roomful of visitors, milling about and making small talk. And dementia can also be responsible for delusions. A friend told me about her widowed mother, who daily met the entire cast of Eastenders lounging about her living room. Is David's God real or not? Nobody knows the truth of these things except God. If He exists, which He might. Remember—He might turn out to be Trev, the spotty-faced guy who takes care of the trolleys at Tesco. All that matters in the current scenario is that David is way too old to survive breaking up with God. Into the lonely God-void, the past rises up and up, until one day his Welsh childhood is more real than the present. What is he doing here in Ohio? It's all been a huge mistake. A series of mistakes and wrong turnings, and oh my—he could cry with wanting to go home.

~~~

'Do you ever think it's odd, to be here? I mean, instead of Ireland?' he asks Margaret one day. He can't quite bring himself to say how homesick for Wales he is. The kitchen is over-hot, with flies circling endlessly and ants marching over countertops. She's shelling peas at the table, and they've not spoken for a while. He's sitting in his armchair by the open window. The same chair they've had for thirty-five years, well patched. Margaret pauses shelling, squints her eyes. Lays down her age-spotted gnarled hands. She's so thin now, she's almost transparent. He can easily imagine her skeleton with the veins and arteries pulsing around it. She's her own gist.

'Well, I reckon you're right, David. It *is* pretty odd, when you think on it.'

'I know! I feel like that too!' he gushes in relief, because even though her voice is cracked and wavery, her old personality is shining through. She's having a good day and they are in attune.

'Nope!' she continues. 'Never expected to end up in a place like this.' Then she sighs. 'I'm just a mite disappointed sometimes. My life is being played out in some kind of…nothing place. The sticks.'

'You're just homesick. We both are, Margaret.'

'What? Not a bit. I just thought I'd end up somewhere more… more exciting. That's all. Maybe New York, with interesting people coming and going. Important things were in store for me, David. That's what my mam always told me.'

David is shocked to see tears on his wife's cheeks. The Margaret he knows is not a crier, but then this is what age can do. Strip away lifelong personalities and inhibitions. And it seems incontinence is not just about bladders, but liquid from all orifices. Oh who cares, he thinks. *Let go.*

'Ah, Margaret.' He sighs, and tosses her a hanky. There's to be no consoling affinity on the subject of homesickness after all, only the task of consoling his disappointed wife. 'Everyone has to live somewhere, love. There are worse places.'

She frowns at him a moment, then blows her nose and out comes her old smile. Her chin dimple is still vaguely sweet, but only because he remembers what she used to look like.

'Yeah, David. Reckon you're right again. Too durn late to do anything about it now anyway. We are where we are.'

They share a glass of home brew, and while not much else of import is said that day, both feel the hard knot of their separate discontents unravel. They've been set against each other for so long, and now they feel a little giddy. David imagines boring down into this semi-gilded moment and staying there a while, at least till God comes back. That would be nice.

~~~

I could console David, because unlike Margaret I share his late-life nostalgia for the place that made me. I may have been compelled to come to Europe by homesickness embedded in my DNA, but now I suffer from homesickness for the Californian suburb of my childhood. It grows as I get older, this ache, and now one of my daughters (made in Scotland) has made her permanent home in California. Almost as if she is genetically answering my homesickness. Which is almost certainly complete nonsense. If anything, she's inherited my restlessness and tendency to imagine the grass is always greener somewhere else. Which in a former incarnation might have turned us into nomads and saved us both from starving. Oh, the consequences of dated biological imperatives! The football games which answer the instinct for territorialism and war. All those people boarding the London tube every morning, stony-faced and dead-eyed but filled with purpose, instead of hunting and gathering. And perhaps all those menopausal divorced women, unaware they're still seeking the best mate to make a viable baby with. It's not love they're after. Perhaps it never was.

~~~

David changes his will three times in the final three years of his life. The first will bequeaths the farm to his elder son William, but then William is kicked by a stallion and dies. The second will favours the next son, Daniel, but then Daniel is put in jail for shooting the sheriff's deputy during a dispute about a woman called Angelina. The deputy is only wounded in his buttocks, and to be fair he'd provoked Daniel by calling him a lily-livered pussy—but still, off the will for Daniel! The daughters are all away in their own lives with husbands and children and grandchildren, aside possibly from the one who's not stayed in touch. The third will, written less than a year before death, is straightforward. *I give, devise and bequeath to my beloved wife Margaret Williams, my farm in Wayne County on which she currently resides.* This will also states four times that his debts be paid from his estate. How extensive these debts are will remain a mystery. I could look, but I don't care to. I'm proud, however, to report that his farm is not sold. Not bad for a runty immigrant kid, to own and successfully farm 163 acres. His parents' dreams have come true. I also note that he describes his wife as beloved. That is not a legal term. He could've just said wife. Proof, if further proof is needed, that theirs is a good marriage overall.

~~~

Forty-eight years after arriving in Ohio, David MacWilliam Williams, age seventy-seven, begins dying. William IV, the Sailor King, is king of Britain. Andrew Jackson, Old Hickory, is president of the United States, of which there are currently twenty-four. It's February. David has chronic obstructive pulmonary disorder, but doesn't know that. He's wheezing in his bed, feverishly waiting for God to speak. To re-join him, at least. A one-sided conversation would do nicely, like the old days. Then, in the nick of time—miracle of miracles! The fifth miracle—God is back in town. No words, but He is close again, and that is all that matters. *Where have You been? Never mind, it's just good You're here now. I've missed you.* David is in the marital bed, the nocturnal separation having ended five years ago. His wife and one of his granddaughters and her children are in the house, but they don't witness his departure. His death, unbeknownst to him, chooses a time when they are all downstairs in the kitchen, drinking coffee and talking low about the practical things that always accompany death. He hears the new puppy whining, and the old dog barks once—probably at the cat. Between wheezes he wonders if they've remembered to feed the animals. He asks God the kinds of things you might ask when you're about to travel somewhere. *Have I got everything I need? Will I need warm clothes? I'm happy with a sofa, don't worry about giving me a bed or changing the sheets.* Everything is fine now. He's never felt so submissive, so trusting, and this may be why his dying is relatively painless and his mood mild. He's not foreseeing an end to the conversation, merely a period of adjustment—and if nothing else, David is very adaptable. Death does not frighten him. Minutes before the end, when his breaths are shallow and few, one of his grandchildren giggles. Sarah Mae is eight years old, red-haired as her great-grandma, and her giggle bursts out of her despite the sombre kitchen. Maybe because of the sombreness, a stance against mortality. It spirals out of the room and up the stairs until it comes to David's ears. He doesn't open his eyes, merely allows the laugh entry. Then he smiles, the smallest quietest of smiles, but still. As if her laughter is the Life force itself. And isn't that fitting, for it carries him off.

Fair winds, David. You've done much, and now you're done.

~~~

After fifty-nine years, Margaret is a widow. Now that he's not physically present to distract and annoy her, and half the livestock have been sold to pay off old gambling debts, David creeps up on her with surprising regularity and vividness. It almost breaks her heart to remember his kind face spattered with freckles, his funny way of laughing. His skinny pale body, bony knees and long bony fingers—how had she ever not noticed how beautiful this man was? For a good three weeks, she fails to recall a single negative thing about him. He was goodness personified. And then, as if there's a David-shaped vacuum in the house now and someone has to fill it, Margaret finds herself praying. Not conversing with God like David, and she's not happy about it either—she hardly feels herself, praying. And yet, here it is, this need to believe in something good. Something safe. Plus it's handy to pass messages on to David via the God conduit. Hey, tell David the harvest is late again, will you? And where did he leave the corkscrew, the good one? And horse manure has rescued those rose bushes he brought me back in '92. Tell him they're covering the porch like crazy.

~~~

One evening, she stands over the kitchen sink and watches the sun sink over the winter-raw fields. Just before it disappears, a gelatinous yellow disc, the sky flares into violet, pink, rose, primrose, turquoise. There's even green. A skein of snow geese rise up in an uprush of beating wings. Whoosh! They circle, mingle and within seconds form into a three V-shapes to cross the sky. It suddenly comes to her that David's God (she cannot stop the habit of thinking of him this way) might be in all of this. In geese and grass blades and infants. Maybe David's God is simply the will to survive in whatever form is required to make continuation possible. Which helps to explain why she's always struggled to get an image of him. Maybe David was wrong— God's not one thing, not a thing at all—he's everything. This idea gives her a sense of expansiveness, of comprehension. But her idea evaporates within the hour. She lacks David's depth and concentration. Or maybe she just lacks the need. What does arrive, and sticks to her, is a nostalgia for her own life—as if she has left it already and is looking back. Or as if she's about to go on a journey, and her past is a country she'll never return to. What an incredible thing it has been—and peopled with such extraordinary people. Oh, she is going to miss it so much! Never seeing this view from her kitchen sink again? Never seeing a son's face again with his crinkly nose, or hearing a daughter's voice explaining again in her exasperated way that the dog has already been fed? Unacceptable, that's what death is.

~~~

Almost three years after her husband died, on a January afternoon, aged seventy-four, my Margaret is dying of undiagnosed uterine cancer. No one knows what to do about her. Some days she's doubled over complaining of stomach pain, other days she's making jam and singing Clementine. It's bleak November and she's still in her prairie farmhouse. The stove keeps the kitchen warm, but it's freezing here in her bedroom. There's ice on the inside of the only window, kept covered with a blanket. Her children are middle-aged now, and two of her granddaughters work at the Florentine Hotel in Germantown. Her daughters take it in turns to swing by, cook a meal, do the washing, make sure the house is warm. And Daniel (married at last!) stops by most afternoons to sit and talk with her. There's been some talk of moving her to a daughter's house, but it's in town and Margaret couldn't bear town life now.

Margaret Beaton arrives one afternoon to find her friend in bed looking dead. She hides her dismay and sits on the bed next to her. Silence, as the two old women consider each other. They've never lied to each other. They acknowledge the proximity of death now with a wry smile, as is their wont. A smile that says: *This is a shitshow, but what can we do?* Margaret Beaton holds her friend's hand, strokes the back of it, then says:

'Magpie, remember the afternoon when we jumped in the river? It was August,' she says in her old lady voice. Wavery and thin, like water.

'Wah.' Which is a word people use these days to mean *Of course. Duh.*

'It was so hot. The kids just stripped off and jumped in. And then, Magpie, you stripped and jumped in too.'

'I never took off my shift.' The words spaced out and almost inaudible, for speaking at all takes enormous effort.

'Oh, believe me, honey—once it was wet, it was invisible.

You were, to all intents and purposes, naked.'

'Stop it! You're a terrible woman altogether, to be telling me a lie like this. Naked, indeed!' The energy required to utter all these words is scraped up from the bottom of her reserves, only accessible because of her righteousness. Her voice sounds transparent.

'I was too shy to take my own dress off, Magpie. I wish I'd been brave like you.'

There's a moment with nothing in it but exhaustion. My Margaret's mouth opens as if she's about to speak, then it closes again in a tight hard line. They squeeze hands and smile at each other. Margaret Beaton begins chatting about the past, the seed potatoes she bought last week, the new great grandchild's strange name, but my Margaret says no more. When she closes her eyes, white lips still vaguely smiling, her friend says the Lord's prayer, not knowing the words for the Last Rites. She keeps repeating this prayer until there is no breath moving in her old friend. Then she makes the sign of the cross on Margaret's forehead, chest, left shoulder, right shoulder. Heats water and tenderly washes her body, dresses her in the navy dress with pale blue polka dots. Anoints her neck and wrists with the oil they made from rose petals a few years back. Kisses her forehead.

'Oh Magpie,' she whispers on and off as she does these things. She tsks and sighs and pines. 'Oh, my marvellous Magpie!'

But there, the tenderness ends. The children are summoned. It's been a bad year, and no one has money for a headstone. They discuss waiting a year or two, jointly paying for a proper headstone for both parents because they are—after all!—good children and responsible citizens. But for now, Margaret is interred on a sleety Friday afternoon with a tiny wooden cross, and within a season there's nothing to indicate which side of David she's lying. Their son William and the infant who was still born are also there, so maybe she's close to them? Daniel persuades his wife Bethany to move into the farmhouse, and while tossing away much of her dead in-laws' possessions, she discovers a gold sovereign sewn into Margaret's oldest corset. My, my! She neglects to tell Daniel, considering it her personal inheritance to be saved for a rainy day. Which arrives in eleven years during the worst drought witnessed for decades. It prevents the youngest Williams from starving to death. The theft that keeps giving.

In sixty-two years, after a prairie wildfire, no one will know where any of the Williams family are buried. Which, sooner or later, is the case for everyone. No matter how courageous and important the life, it will be forgotten—and surprisingly soon. Oh, Ozymandias! And that is the end of the story of Margaret Lyons and David McWilliams Williams, resurrected for this narrative and then laid to rest again. Or is it? Is there ever a final ending? Maybe we circle round and round inside our narratives, our one-year-old self neatly tucked into the centre.

~~~

Back to the birthday boy, with his whole life ahead of him. That unlovable child back in Llangain, 4000 miles to the east of Ohio. Freckles like blackcurrants, skin like non-fat milk. I'm worried about him. Aren't you?

'David,' I say.

'What.'

'I just want to say that, well, that basically everything is going to be alright.'

'Are you insane?'

'A little.'

This might be my last ancestor visit for a while. I'm exhausted with the complexity of lives, the sorrow and pain, even the exuberance. I'm almost seventy, and energy for socialising is starting to wane.

'Are you sleepy? Maybe you need a nap,' suggests David, and then commences sucking his thumb as if to demonstrate the proper soporific pose.

'You're right. I'm running out of steam. It's not that I don't want to spend more time with you, but…'

'You want to break up with me. With all of us.'

'No! Never! Well, kind of. Maybe a trial separation. Here. Choose a hand.' I hold my two closed fists in front of him.

'Oh! I love games!'

I'm starting to think he's the sweetest natured ancestor I've met so far. Not an ounce of crankiness. And I started out thinking he was so runty, so sour smelling and runny nosed. What else have I been wrong about? He chooses my left hand and I open it.

'It's empty!' Tears spring into his eyes, which I now see as beautifully intelligent.

'Here you are!'

I quickly open my right hand to reveal a small parcel in shiny gold paper. He rips it open and pulls out a small ship in a bottle. (I've been here a few times and given him a few gifts—none of which he remembers. This time I've had a gift custom-made. I'm really excited about it.)

'Oh!' he cries. 'Oh! Oh! What's her name?'

'I don't know. She's your sailing ship, you choose a name.'

I'm not sure I've ever given a gift so appreciated. This child is incandescent with joy. Then he drops it on the stone floor, the bottle shatters into a hundred pieces and the ship into dozens of splinters. He makes no noise, just stands there mouth open, a quizzical look on his face. Like—this cannot just have happened. Also like—of course something like this was going to happen.

'Can you fix it?' he whispers.

'No,' I whisper back.

He sighs, closes his eyes for a moment, and when he opens

them again he's calm, stoical—which reminds me of William the worm, and a dozen other ancestors including my mother when I let her down and she forgave me again. Then without a word or glance, just one hand briefly flicked up as he turns his back on me—a farewell? a thanks-anyway gesture?—he toddles quietly back to his dark corner in the kitchen-living-room-bedroom. Not all birthday visits are successful, but it's especially anticlimactic for the final one to end on this note. Or maybe it's apt. Maybe there's no such thing as a good or bad ending, any more than there is good or bad weather. It's weather! It's Life! Thomas Hardy says in the final lines of The Mayor of Casterbridge, *Happiness is but the occasional episode in a general drama of pain.* That probably shouldn't make me smile but it always does.

Finale

Happy birthday Life
Everywhere, all the time

It's a one man show for everyone, no matter how married or surrounded by friends and children we are, and whether we like it or not. We guess what others think and feel—indeed, how they perceive the world. Do you feel the same way I do, for instance, upon entering Tesco at midnight and seeing all the foreign workers stocking shelves and none of them are people you've seen around town during the day? Or seeing a child, a stranger, give you a cheeky wave out of a passing train window? There's no way to know how others really feel, beyond the physical signals we've all learned to exhibit in certain situations. Yet simultaneously, it's also a family show. Imagine viewing the history of life on Earth via a hyper-fast speed film, keeping in mind nothing is added or removed. Watch it. It looks like one thing that keeps reinventing itself. A nebulous living form that takes on alternating shapes—both individual and social. There it is—wriggling, morphing, rising, melting, soaring again briefly. *There we are*. A single thing.

We began with a life form called Kevin, and we're all verifiably related to each other. It gets cosier when I consider I have roughly 15,000 distant cousins alive *right now*. Lots of them are 25th or 667th cousins. Only thirteen first cousins, but still. 15,000 cousins. (Are you one of them? Come for Christmas!) On the other hand, there's almost eight billion humans alive right now, which sounds a lot—but it's not. We only make up .01% of all life on Earth. As a species, we suffer from delusions of grandeur. Perhaps this is accounted for by our intrinsic optimism, not hubris. Or maybe that idea is yet another positive spin I've given myself. Why do we assume we're so special? 99.9% of the species who've existed are already extinct. The average life span of a species is four million years, and we've only been around 300,000 years. On paper, ultimately our chances don't look good. We have time, but not endless amounts. Maybe all life forms are programmed to evolve until they are no longer effective Life citizens, whereupon they begin the descent to extinction.

~~~

And I keep coming back to this: Maybe Time is not how we perceive it. After all, there's no other word in English for time but time, which might indicate we're not confident enough to give it variations. Not only is our perception of its passage relative to our situation (thank you Einstein), from a physics point of view, the fact we experience it chronologically is no proof against the past, present and future happening simultaneously or even continuously. Australian Aboriginals, among other cultures, think that temporality and eternity are not separate. That we *were*, *are* and *will be* in a state the Aboriginals call Dreaming—which is also our perceived life. Maybe the passage of time is not along a horizontal line but vertical, and the past and present happen simultaneously. Or maybe time tunes into emotion, bends to it, breaks the rules for strong feelings. Think of *An Occurrence at Owl Creek Bridge* by Ambrose Bierce. The moments between a noose being hung around a man's neck and the rope breaking his neck as he falls—this ellipse of time contains a whole reality, including an imagined escape and homecoming. Scottish author Andrew Greig once said: 'I have a notion, which could be sentimental delusion but which nevertheless haunts me, that there is a place where our most intense moments continue to exist. Surely such intensity, such clarity burns itself through one layer of being into another.' Nobody knows about Time. Nobody knows much about anything with certainty, which to my mind is exciting.

~~~

Maybe all the things that made me are floating behind and above me right now—including Kevin, Polly, William, Harriet, Cora, Florence, Polly, Elizabeth, Hans, Betty, Norma, Walking-Fish-Margaret, Susan, Freddie Tony, Jeff, poor poisoned Margaret Drummond, Jeanne, William. Maybe death marks the end of the tyranny of time, and my beautiful mother is aged twenty-one with her red lipstick smile and legs that work. She's thinking about that new man in the office, the guy called Micky—my father. He's aged twenty-three in his first gold corduroy jacket, walking down Market Street with the sun glinting on his blond hair. Their parents and their parents, on and on. Maybe they are all still in their primes, moving with me whenever I move like a murmuration of swallows. 'We hold in our energy systems people we have never met,' according to author Kapka Kassabova in *To the Lake,* referring to the unknowable influence of ancestors. The world is jam-packed with life forms in a zillion different shapes and sizes, every one of them a re-mix of the same ingredients that have always been here. And just around the corner, no doubt, are all the living things that will result from a re-mix of us.

~~~

One day, in a post peanut butter cookie mood (try it), sluggishly reckless, I close my eyes, count to three, then keep going. *Whoohoo!* Whizzing past my genesis, past the ancestors, past eons. I'm in a desiccated Kenya, I'm in a jungle, I'm in a place ripped asunder by an earthquake. Everything is dark, everything is light. It's freezing, it's hot, it's cold again. Then I'm in the sea, and I'm in the sea, and oh look—I'm still in the sea. I'm in the air, and it has no oxygen. Suddenly—*wham!* Who's first birthday is it today? Well, it seems to be everyone's. Look around. They're all here. (*We're* all here.) Homo sapiens comprise a teeny fraction of the birthday children. Less than a hundredth of one percent, though naturally I gravitate to them because I'm shy and feel more comfortable with my own kind. We form a mini ghetto at the outskirts of what is beginning to feel like a party.

'Hey Dad! Is that really you! Love the corduroy jacket. Can I borrow it?'

'No.'

And oh my God, is that Margaret Drummond from the sixteenth century?

'How's it going,' I say, and give her a hug because by now I feel we're old friends. She knows me this time, which is a first.

Then up walks tsunami-survivor Florence, her wild red hair still a mess but gorgeous now, like it's a style—and I sense it's alright to embrace her too. She knows the score.

No one is living through chronological time here. Not on this birthday. The air is thick with beings whose hearts no longer beat—and yet I can hear our pulses, the thud-thuds, the ba-boom ba-booms, the tinny tick tocks. We are all babies, of course—a one-year-old is definitely still a baby—and yet, no one here is an infant. We can all sing and talk and run and dance like grownups or the finest grown-up version of our species. Audacious, deluded, amateurish, ridiculous us! Then it hits me, like David William's epiphany in the cow shed. We seem infantile because we *are* infants. Always have been, always will be about one year old. The place is dotted with birthday cakes and a quadrillion versions of cakes. Some look like rabbit poops, but who am I to judge? Today is a hundred birthdays, a thousand, a hundred thousand. Oh much more than that. The true number would fill this book with zeros, and the book itself would be too big to print.

~~~

'Jesus! Can we please dance now?' asks Jeanne Breconnier impatiently, her bosom heaving in her yellow slightly damp-smelling King's Bride gown, her slipper-clad feet tapping.

'Uh, sure,' I say. Now the idea of dancing has sprung up, there's suddenly music—from my era, which no one thinks is weird but me. I hear Hound Dog playing somewhere, then Footloose, then Sugar Magnolia. Where is Elizabeth, my bubbly hominid grandmother who had a momentary lapse of sense with a Neanderthal called Ned? She's the best ancestor for dancing, surely. And when I spot her, she's eighteen and dancing with Tony! Oh my God, how has she coaxed shy melancholic Tony on to the dancefloor? And where is his wife Luciana? Yikes! She's cavorting with William the worm!!! An inter-species dalliance, but they are easily the handsomest couple in the room. I wave to them all. Elizabeth doesn't see me but William waves as much as a worm can, and smiles his slow Eeyore smile.

'Come on over! Come dance with us!' he calls.

But I'm more like my dad, not a graceful dancer, and in any case I'm afraid to walk to the dance place in case I step on a Kevin, for the ground is moist and alive-looking. The noise becomes a cacophony, like it does after a while if it's a good party. I can't hear any individual sound, just a single humming roar—like when my youngest grandchild mixes all the paints up until there is just brown. Sound is only meaningful when there are pieces of silence around it. Currently at this party, there is no silence around sound. I cup my hands in front of my mouth, and from there forms a visible tunnel of silence into which I whisper:

'Kevin. Kevin! Are you here?'

Nothing. What did I expect? He's probably smeared like gum on the sole of my shoe.

Then a tiny voice. I can't see him, but I speak in the direction of his voice.

'Are you really here, Kevin?'

'Of course I'm here. I've been waiting for ages.'

'You have?'

'I just told you that, didn't I?'

'Huh. Sorry for keeping you waiting. I'm here now.'

'You're so vain. I'm not waiting for you, I'm waiting for the toast. You said you toasted me nightly, to thank me. *I raise a glass to Kevin,* was your exact phrase.'

'Ah. The preface of this book. That was a, a, a manner of speaking. I liked the Keat's post-humous quote.'

A silence. Is Kevin pouting?

'Well, anyway,' he finally says in his thin reedy voice. 'Now you're here, pour the damn drink.'

And so I do. I open the bottle of champagne I suddenly find in my left hand, nicely cooled and de-corked. I pour a glass for myself and carefully fill the tiny shallow dish I find by my right foot. Then I lift my glass. It feels a grand gesture and I stand there a minute, posed for God knows what. Am I expecting someone to take my picture? The tunnel of silence between me and Kevin continues and no one at the party notices us. Beings are still filing in. The place is so packed, it's hard to distinguish individuals. I can see everyone's mouth moving and laughing and they are all, depending on species, waving or dancing or wriggling or squirming or sliding or swimming or flying. One of the Georges is falling over, and Freddie is hitting on Fiona like a hopeful adolescent. Some of the Margarets are meandering near the buffet, empty plates in hand. I find something to stand on, pray it's not a life form, and raise my glass higher yet. With all the dignity I can muster under the circumstances, I cough and say:

'Here's to you, Kevin. Thank you for my life.' I space the words out solemnly to make sure he hears.

'You are welcome,' he says, equally solemnly. I picture his unicellular body giving a bow, and his version of a mouth smiling. How do I know he's happy? I know this about my ancestors: They love to drink and they love attention. I take a sip and the tunnel of silence evaporates. My ears fill with the music of countless beings celebrating the fact they're alive. There's no gap, but that's okay now. I briefly miss the private moment I had with Kevin. Then I turn to the gathering and think the word *Quiet!* Everyone quietens and looks at me, like *Who the hell are you*. I raise my glass one more time and say out loud:

'Here's to you all.'

And bingo. Glasses appear in all their hands, or versions of glasses in versions of hands.

'Here's to us all!' an uncountable number and versions of voices say.

'Thank you,' I add, but I don't think anyone hears me anymore. That's okay.

~~~

Parties take a lot of work and sometimes they flop because the thing that makes a party take off can't be orchestrated. But my ancestor party isn't flopping. A delightful madness is spreading to every life form. Much foolishness and indiscretion, much to regret tomorrow. It's uncanny how realistic the party looks, how permanent feeling. Voila! Here we all are, I think. (Or rather, here I am. You might be anywhere.) And then, just when I can't imagine it ever ending, it's over. Off we all go. When I look over my shoulder a minute later, it's like we never happened at all. Which is, actually, much more likely than us happening in the first place. I remember how Earth began. The hungover woman waking up and trying to remember who and where she is. It's pretty random. She's no one special; last night was not special. She's thinking: This was not the plan for today. Tiptoeing around, slowly gathering the things she needs and hunkering down with them, while being hit on the head over and over by asteroids. Not to mention erupting volcanoes. Waiting for what seems an eternity, not even knowing what she's waiting for. We are waiting offstage. That's us, in the shadowy wings. Un-auditioned actors, pulses of potential. Hopeful chancers every one of us, waiting without knowing we're waiting for a crack at existence.

Acknowledgements

Thank you to Doctor Andrew Easton, professor emeritus at Warwick University, for being my science fact-checker. Thank you to Anne MacLeod for pointing out that I was really writing two books, not one, and travel devices like egg timers were probably not needed. Betty Vaillant helped with my non-existent French, pointing out when my Parisian character was actually speaking Italian. My cousin Linda Engbrenghof offered endless help with family history. I'm very grateful to the wisdom of Adam George Fletcher for his perspective on the brevity of our life spans. Not bad for a nine year old. And I owe an enormous debt to Graham Bullen and Cait McCullagh, both of whom offered me invaluable detailed feedback on style and content. Thanks to Peter Whiteley for his assiduous proofreading and generous praise of my puns – all of them accidental, but no matter. I'm very grateful to Sparsile for taking a whirl with me, and in particular, Lesley for creating the best cover I've ever had.

My references
(in addition to articles online and in the Guardian):

The Periodic Table by Primo Levi
A Short History of Nearly Everything by Bill Bryson
The Body—A Guide for Occupants by Bill Bryson
The Selfish Gene by Richard Dawkin
Sapiens—A Brief History of Humankind by Yuval Noah Harari
The Story of Life by Catherine Barr and Steve Williams
On the Origin of the Species by Charles Darwin
Billions of Years of Amazing Changes—the Story of Evolution by Laurence Pringle

Evolutions—15 Myths that Explain the World by Oren Harman

The Kings and Queens of England and Scotland by Plantagenet Somerset Fry

To the Lake by Kapka Kassabova

Note: I created almost all the ancestors in this book. However, Margaret Drummond, Jeanne Breconnier, Margaret Lyons and David Williams are verifiably my distant grandparents. One Margaret really did have a baby with the king of Scotland and die of poisoning along with her sisters, and the other Margaret really did wash George Washington's dirty underwear at Valley Forge and marry at age 13. Jeanne Breconnier really did travel from Paris as a King's Daughter to marry three strangers in two years in Canada. Richard Swarts really did impregnate my teenage grandmother twice before hopping on a freight train. And David Williams really was a revolutionary solider and is recorded as one of three to survive a shipwreck.